EMS SAFETY

Techniques and Applications

Federal Emergency Management Agency

United States Fire Administration

Federal Emergency Management Agency
United States Fire Administration
Emmitsburg, Maryland 21727

This manual on Emergency Medical Services Safety has been developed to assist EMS providers in the reduction of line-of-duty injuries, illnesses and fatalities. It provides a framework for developing programs that will create an appropriate margin of health and safety during the performance of their duties.

The hazards of hostile or unknown environments, communicable diseases, violence, hazardous materials and critical incident stress unfortunately continue to be on the rise and effect those emergency responders in the fire and EMS departments across the United States. Additional hazards such as those associated with driving, lifting and carrying, and overexertion also effect the morbidity of those providing EMS services.

This manual is intended to be used for the implementation of programs that identify the hazards faced by EMS providers and describes the ways in which these responders can and should deal with these risks.

We encourage all EMS providers to address the issue of occupational safety and health programs. A properly protected, trained and staffed EMS response will not only provide a greater margin of safety to the members, but will enhance the service to the citizens that we are sworn to protect. With a positive approach to these issues we will better understand and, most importantly, prevent the injuries and fatalities that plague this occupation.

We would like to thank the fire fighters, paramedics and emergency medical technicians and their respective organizations for assisting in the development of this manual. The Federal Emergency Management Agency, United States Fire Administration is proud to provide this first comprehensive manual that seeks to protect the health and safety of all EMS providers.

Preface

The United States Fire Administration (USFA) publication EMS Safety: Techniques & Applications was developed as a comprehensive manual to address the hazards faced by EMS providers and describe ways in which emergency responders can and should deal with these risks. The purpose of the manual is to reduce the number of EMS personnel killed and injured in the line of duty across the country by providing them with the information they need to create an appropriate margin of safety during the performance of their duties.

This manual addresses a broad range of safety considerations for firefighters and other EMS providers. Topics addressed include the universal elements of safe EMS operations such as vehicle operation and personal protective equipment. Scene operations, health maintenance, and manager responsibilities to safety are also included in the context of this manual.

EMS Safety: Techniques & Applications was developed by the International Association of Fire Fighters (IAFF) under Federal Emergency Management Agency contract #EMW-91-C-3592 for the United States Fire Administration. Principle contract personnel were Richard Duffy, Director, Occupational Health and Safety Department, IAFF; Sharon Doyle, M.P.H., Occupational Safety and Health Assistant, IAFF; Tom Scott, Director, Emergency Care Information Center (ECIC), Jems Communications (JEMS); Kate Dernocoeur, Technical Writer, ECIC, JEMS; and Elizabeth DiPinto, Design and Production, ECIC, JEMS.

An advisory committee was selected based on an extensive range of expertise. Advisory committee members included: William H. Leonard, EMS safety expert; Paul Maniscalco, National Association of EMTs; Jim Melius, M.D., Dr.PH., New York State Department of Health; Carlos Perez, Office of Trauma Services, Metro-Dade Fire Department; Ameen I. Ramzy, M.D., National Association of State EMS Directors; Susan Ryan, National Highway Traffic Safety Administration; Gordon M. Sachs, U.S. Fire Administration; Charles J. Simpson, Jr., National Volunteer Fire Council; Ed Stinette, International Association of Fire Chiefs; Bruce Teele, Senior Fire Service Safety and Health Specialist, National Fire Protection Association.

Review committee members provided invaluable comments on the written drafts of this manual. Review committee members also were selected based on an extensive range of expertise. Review committee members included: Steven A. Fort-y, Ambulance Insurance Services; Kevin McGinnis, ASTM F-30 Committee on EMS; Michael McKitrick, Chicago Fire Fighters Union, IAFF Local 2; Ron Myers, American Ambulance Association; Michael O'Keefe, National Council of State EMS Training Coordinators; John Sinclair, Central Pierce County (WA) Fire Department; and Bill Stevenson, Sussex County (DE) Paramedic Services. Various other individuals also lent their time and knowledge in reviewing specific chapters of this manual.

Other USFA publications and National Fire Academy (NFA) courses address specific aspects of EMS health and safety Many of these are listed in the reference sections of chapters in this manual. For a catalog describing USFA publications, or for a catalog of NFA training activities, please write USFA publications, 16825 South Seton Avenue, Emmitsburg, Maryland 21727.

Contents

SECTION I
Universal Elements of Safety

CHAPTER 1

Why Safety?

Chapter Overview: Prehospital care providers must make numerous decisions on each call. Some of these decisions relate to the constantly underlying theme of safety. This chapter provides an introduction to this comprehensive safety manual. Topics addressed range from immediate threats (from the prehospital environment or the people in it) to long term threats related to the lifestyle commonly associated with EMS and stress. This chapter also presents a discussion on the nature of crisis, which helps contribute to the unpredictable and sometimes volatile realm of EMS.

The number of decisions prehospital emergency medical personnel make on every response is impressive. These decisions may include selecting the best route to the scene, negotiating traffic, choosing the best interpersonal approach with strangers in crisis, determining needed medical care, and concluding the response with appropriate transportation of the patient.

Safety should always be a primary consideration when making these decisions. For example, the smart EMS provider is always thinking, "How can I negotiate this traffic safely?" Or "How can I work with these people safely?" Or "What about this scene needs to be moved or changed so 1 can work safely?"

Safety is an underlying theme that no one can afford to forget. Yet, it's because this vital element of EMS is such a constant issue that it sometimes becomes a backdrop rather than a priority

The delivery of emergency medical services (EMS) involves hazardous work. Anyone who has spent time providing emergency medical care knows this. Consider the headlines that appear all too frequently: "Traffic fatalities involving EMS units. " "Downed power lines hamper rescue effort." " Paramedics held at gunpoint during narcotics robbery attempt." "EMT struck by hit-and-run motorist, critically injured." "Dogs attack would-be helpers." "Firefighter succumbs to hepatitis." "Retired EMT dies of heart attack. . . stroke . . . cancer. . ."

The same safety issues are common to all EMS personnel. They include physical or environmental hazards, traffic, the frequently strenuous physical labor and emotionally charged interactions. Safety issues cross all levels of certification, territorial boundaries and population demographics. Yet so little emphasis is placed on workplace safety for EMS personnel that as of January 1992, there was little research about safety, with statistics remaining largely unavailable in EMS literature.

With the advent of modern emergency medical services in some cities in the early 1970s sophisticated equipment and education met an important social need. Two decades later, this medical specialty has earned wide recognition and increased legitimacy within the medical community The days of fighting for survival and acceptance have passed. Now, refinements to the basic structure of EMS systems can begin to receive overdue attention. The issue of safety-which in an ideal world would have been first on everyone's list-is increasingly highlighted.

This manual is dedicated solely and comprehensively to EMS safety For field personnel who have always felt the need to address personal safety it is a welcome addition to the literature. For chief officers, managers and agency executives who have compelling reasons of their own to focus on safety, this book can serve as a worthy guide.

Supporters of a safer workplace- on both sides of the administrative fence-have historically lacked a solid foundation of good data upon which they could construct a safer workplace. Instead, those inclined to focus upon safety have usually relied on assumptions and the lessons of personal experience. A national trend toward promoting research forums and foundations demonstrates a sincere effort to fill this vacuum of safety information and statistics. The intended result is improved prehospital safety.

A major safety factor for EMS personnel is their degree of vulnerability. Every day, EMS personnel violate two of the basic safety messages learned in childhood: "Don't play in traffic" and "Don't talk to strangers." When something goes wrong, the resulting injuries or illnesses are seldom as simple as what might befall office workers or even health-care providers who work in more controlled settings.

The unpredictable nature of prehospital care attracts adventuresome people who wonder, "Could I do that sort of work?" But it also eventually scares many of them away, once they have some personal experience with the risks. Fear-based attrition-an unresearched but anecdotally real phenomenon-is unfortunate and largely unnecessary Most of the risks associated with EMS can be predicted and either prevented or minimized. Each individual crew must apply the principles of safety and use preventative tactics; and managers must support and promote safety-consciousness. Proper attitude and a commitment to safety set the groundwork for this. Unless both people in command or leadership positions and those actually in the arena of prehospital care regard the issue of safety seriously and genuinely, safety programs are likely to be less effective, with potentially tragic results.

The principles outlined in this manual are wide-ranging, including how to approach scenes safely, principles of safe driving, use of seatbelts in the cab and the patient compartment, interfacing safely with hazardous material teams, safe interpersonal communication strategies, back safety and even the long-term safety-related views of health maintenance through physical fitness and stress management. Although it is reassuring to know loved ones will have a financial cushion in the months following the death or permanent disablement of an EMS provider: the better answer is to avoid death or injury in the first place.

The Nature of Crisis

A critical safety concern in EMS is the way people respond to crisis. People at the scene of a medical emergency are frequently dangerous. Often, even at "harmless" scenes, neither the lay public nor EMS personnel recognize the impact of their interpersonal relations and individual reactions to an emergency. A crisis can make normally placid and cooperative citizens disruptive and even violent. What follows are the important principles about the nature of crisis:

• The effect of adrenaline has a great impact on people unaccustomed to emergencies. People often overreact to such situations because of a surge of adrenaline. The body's natural response to any perceived crisis is to release this "fight or flight" hormone. It generates increased heart rate, faster breathing, sweaty palms and oftentimes explosive energy.

After working awhile in emergency services, EMS personnel become desensitized. Their bodies either learn to wait until the situation can be assessed before responding with adrenaline, or they become accustomed to the feeling. The difference between what is happening, physiologically, to the adrenaline-charged patients and bystanders and the desensitized emergency responders can cause problems in interpersonal communication. In fact, many of the interpersonal hazards EMS personnel report are due to lack of rapport. Avoiding this means remembering at all times to try to see things from the citizens' point of view; had the bystanders been able to cope, they would not have had to call 911 in the first place.

• *Crisis situations are emotionally charged.* Intense emotions are often evident in people involved in emergency situations, and they can elevate the danger of a situation. The circumstances surrounding a call for help can also be a contributing factor. Consider that adrenaline is surging and the emotions of the people at the scene are intensified. The emotions tend to be negative: anger, frustration, anxiety Often, the outlet for these emotions is violent. Those involved in a neighborhood dispute often remain angry once the fighting has ceased. Although violence may be projected toward arriving emergency personnel, the root of these negative emotions is usually from another cause. For example, one man was furious at himself after he failed to put up the safety gate and his toddler fell down the stairs, His anger was initially directed at the arriving paramedics, who defused it appropriately.

Another nearly universal emotion is fear: of dying, of the unknown, of needles, of the costs involved, of loss of control. Frightened people behave unpredictably Like a drowning person struggles to climb onto would-be rescuers, frightened people sometimes grab and hold tight to EMS personnel. Being grabbed is unsafe, whether the act stems from panic or aggression.

• *Human beings generally do not like change.* People are creatures of routine and habit. Change is disruptive, so it is natural to resist it. When emergencies occur, people often react negatively (sometimes against the first convenient source-which may be EMS personnel) because each emergency is an unplanned change. This element of crisis should be kept in mind even when responding to a nursing home for what the EMS provider may consider a routine transfer. There are all types of disgruntled people; a bite or a kick may be the only way some people can express their frustration at not being able to control their own lives.

The public still needs education about the EMS mission. There will always be people who do not understand proper utilization of local emergency services. Drivers are inconsistently cooperative about helping EMS vehicles pass safely. Many people are unaware of basic anatomy and physiology; one young woman angrily told a paramedic to just "give me a pill that will fix it now!" for her two obviously broken legs. Hostility between EMS personnel and consumers who misuse and abuse the EMS system is common. Public education about EMS must be ongoing and community-wide; in the meantime, safety issues will continue to arise.

One enduring misperception by the lay public about prehospital care is that emergencies are always a matter of life or death. The reality is that for every truly life-threatening situation, there are usually dozens of other prehospital situations that could be termed "low-level" emergencies or even "routine." This can generate dangerous attitudes among EMS personnel (see Chapter 4). And even though a situation seems minor medically, EMS personnel may still face serious safety issues. One prehospital crew was transporting with lights and siren a patient whose chief complaint was a sprained ankle. The ambulance hit a pickup truck. An N-year-old honors student was physically and mentally disabled for life.[2]

Summary

The safety issues addressed in this manual are wide-ranging. The purpose of this chapter is to introduce the reasons why safety is a priority, what some of the issues are, and how the nature of crisis plays a role in the topic of EMS provider safety. Causes of safety issues are diverse. At times, the focus concerns a specific potential risk; other times, the principles are very generalized. This manual is intended to address safety along an entire spectrum, from "micro" to "macro" points of view.

References/Endnotes

1. In 1976, the United States government enacted the Federal Death and Disability Act. In 1986, it was amended to include public-service EMS providers. The Public Safety Officers' Benefits Act (P.L. 94-430) was further amended in 1989. Now, EMTs and paramedics can receive death and disability benefits on par with other public officers. (Those working for commercial or "private" ambulance companies are excluded.)
2. The Pantograph, Bloomington, Illinois, December 29, 1989 (page 1). This tragedy will cost that city $4.975 million if the student lives to her expected life span.

CHAPTER 2

Emergency Medical Vehicle Operations

Chapter Overview: Just because an emergency medical vehicle is intended for honorable purposes, it is not a "magic box" of protection It is a moving vehicle, making those inside vulnerable to injury. This chapter describes elements of vehicle safety related to riding in the EMS unit (both in the cab and the patient compartment) as well as general principles of safe driving. (Safe roadway operations are described in Chapter 7: Outdoor Operations.)

Americans drove more than 2 trillion miles in 1989. There are nearly 195,000,000 registered vehicles in the United States. There is a good reason why parents tell their children not to play in traffic. Roadways can be a dangerous realm for EMS personnel, who spend a good deal of time driving-often in dismal conditions.

Statistics from the fire service alone indicate that more than 6,000 firefighters in the United States were injured in 1989 on the way to or from the scene of an emergency.[1] Add the numbers associated with private and third-service EMS organizations, and a significant safety issue emerges. Anecdotally, many of the injuries sustained by EMS personnel are related to motor vehicles.

Many EMS personnel share an illusion of invincibility. Evidence of this is seen in media photographs, in which other emergency workers are wearing protective gear and EMS personnel are not. One source described this sense of invulnerability as the "Magic Box syndrome,"* in which the overriding attitude among too many EMS personnel is that "it can't happen to me." For example, some fail to consider the hazards within the patient compartment. They regard it as a place as safe as their own living rooms. Others have not had proper driver training. They learned through hands-on practice on the job. With luck, they did not learn too many of their preceptors' bad habits.

Driver training for emergency vehicles is now widely available and should be part of every EMS provider's training. Several nationally recognized programs are available. The purpose and intent of any well-regarded program is to minimize injury, death and damage to expensive equipment. A parallel decrease in risk of liability is likely to occur as well for agencies that provide emergency vehicle driver training.

Instruction should include all aspects of both non-emergency and emergency vehicle operation, including use of the emergency warning systems, communications system, vehicle locating system, on-board computer system, location of area emergency facilities, proper parking and backing procedures and safe driving practices. Also, awareness of the liabilities inherent in emergency vehicle operation must be properly instilled. A new driver should have the assistance of a trained driving instructor for a certain number of initial miles of driving, in both emergency and non-emergency modes.

Many different types of vehicles are used to deliver EMS. Among ambulances, there are three federal classifications (see Figures 2-1 through 2-3). Other emergency medical response vehicles include rescue trucks with specialty equipment, crash trucks, disaster command post vehicles, supervisor vehicles, quick-response-team vehicles and one-person response units that contain advanced life support equipment but have no patient-transport capability.

One effort to minimize the hazards associated with EMS units has been the establishment of national standards for EMS vehicles. ASTM F1230-89, Standard Specifications for Minimum Performance Requirements for EMS Ground Vehicles (a document based on voluntary consensus) and

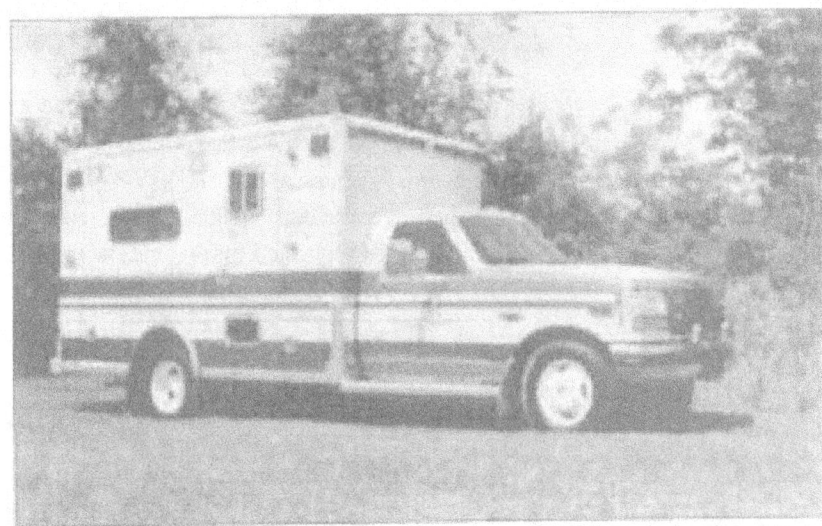

Figure 2-1: A Type I emergency vehicle has a conventional pick-up truck cab and chassis with a modular patient compartment.

Figure 2-2: A Type II emergency vehicle has a van-type configuration with a raised roof.

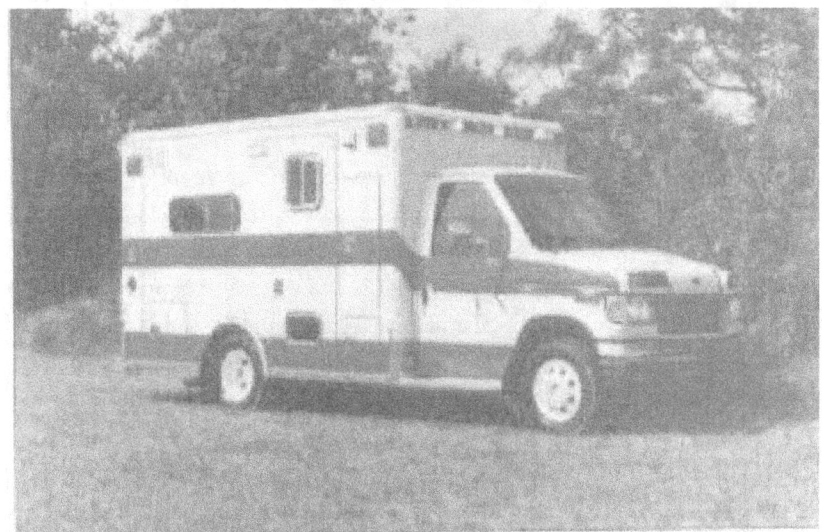

Figure 2-3: A Type III emergency vehicle has a conventional van cab and chassis with a modular patient compartment.

KKK-A-1822C, the set of specifications for ambulances purchased by the federal government, are the two major standard-setting documents. For example, the weight of a vehicle has bearing on its safe use. Government specifications stipulate that ambulance vehicles should not exceed a certain gross vehicle weight (GVW). Compare the maximum allowed GVW to their average curb weight as they leave the manufacturers/(after the special features requested by purchasers are added):

Ambulance Type	Maximum GVW	Manufacturer's Curb Weight
Type I	10,500-11,000 lbs.	8,630 lbs.
Type II	9,000 lbs.	7,065 lbs.
Type III	10,500-11,000 lbs.	8,860 lbs.

Curb weights increase once these vehicles are equipped. Add people, and some will exceed GVW This affects handling and braking. It may not be a coincidence that the highest percentage of EMS unit accidents-between 50 and 70 percent-occur at intersections, which involve sometimes-rapid deceleration.[3,4]

When possible, situating equipment should be done with the users in mind. Driving is safer when the vehicle operator can reach for switches and radio equipment without looking down or reaching beyond normal range. Hands-free radio technology is also now available; anecdotes are ample about avoidable collisions that occurred when radio cords got tangled in the steering wheel.

Part of emergency vehicle safety lies with those with whom EMS personnel share the roadways. Most lay drivers are responsible and well-intentioned, but it takes only one bad or inconsiderate driver to generate a hazardous situation. Although the numbers of people driving while intoxicated steadily decreased in the 1980s alcohol impairment remains a common contributing cause of highway problems; of drivers killed in crashes, 55 percent were intoxicated.[5] It is easy to extrapolate that increased exposure to traffic places EMS personnel at proportionately higher risk for injury and death than people who drive less. No reliable or comprehensive statistics are available to demonstrate this risk because there is as yet no standardized system of reporting EMS traffic mishaps. However, most EMS personnel can relate anecdotal experiences in which they were nearly hit by another vehicle.

One large-scale study examined all crashes involving greater than $600 damage to EMS units in the state of New York in the 48-month period between January 1, 1984, and December 31, 1987. It showed that EMS unit crashes do not necessarily occur in bad weather or poor visibility, on dark roads or at dusk, on wet or snowy roads or while passing a vehicle that refuses to yield. The weather was clear 56 percent of the time; it was daylight 70 percent of the time and dry 63 percent of the time. Only 21 percent of crashes were sideswipes while overtaking.[6]

In The Cab

Most EMS units have seats for two people in the cab. Safety principles associated with the cab include:

• *Always wear the lap and shoulder belts.* The best reason is the improved chance of survival in a crash. In addition, safety restraints help a person stay in position when the forces of momentum threaten displacement. This is especially relevant for the driver; maintaining good position behind the wheel may allow him to continue to maneuver the vehicle, reducing the chances of a collision. Proper posture while sitting can reduce tension and fatigue, which can also help minimize chances of crashing. Driver-side and passenger-side airbags (when available) should be included in specifications for new ambulances to increase safety in the cab.

• *Do not leave loose items on the dash.* Many EMS personnel without a "home base" spend their shifts in the EMS unit. During lulls between calls, they may read, listen to music or even watch a small, portable television. Loose items such as books, electronic appliances, clipboards, food and drink are often placed on the dash. They seem harmless-until the EMS unit is subjected to unexpected changes in momentum either during emergency response or while driving normally. Those loose items can then become harmful missiles.

• *Use caution when exiting the EMS unit.* Injuries are sometimes sustained during the moment of distraction that occurs while arriving at the scene. The EMS crew can finally see the scene and is eager to assist the patient(s). Focusing on patient care logistics may prematurely overtake the need to remain aware of traffic safety while exiting the vehicle. In order to exit the EMS unit safely, EMS personnel on both sides of the EMS vehicle should check the rear-view mirror for approaching vehicles, bikes and people before getting out. Since there is no side mirror for those exiting the patient compartment from the side, the door should be opened slowly, and just enough to be able to see oncoming hazards before getting out. (see Figure 2-4) (Associated safety issues, addressed elsewhere in the text, include quickly surveying the scene for downed power lines, hazardous materials and obviously unruly persons.)

• *Both occupants of the cab should always watch the road.* One day, a paramedic was looking down at his shirt, trying to scrub some blood off his sleeve while his partner drove to an emergency call. Another car hit the ambulance, causing it to roll over. That paramedic did not see the circumstances surrounding the moment of impact to support his partner's report that the ambulance had the green light. Another paramedic used to read a book while her partner drove to EMS scenes. The ambulance was their "office" for each 10-hour shift. An attitude of indifference stemming from familiarity, combined with the "Magic Box syndrome," spawned a cavalier attitude. One day, the paramedic suddenly realized that if the ambulance crashed, there would be no way to honestly report the facts. Furthermore, the driver had no assistance in clearing intersections and other road hazards. From then on, the book was stored in the book bag behind the seat when the ambulance was moving, and a safer habit was established. Both partners should be awake when the EMS vehicle is in motion--even late at night on 24-hour shifts (see Figure 2-5).

If the driver requests, the person in the passenger seat in the cab can help by saying "clear right" to confirm clear intersections. In addition, the partners should share information as they arrive at the scene. For example, when arriving at a multi-vehicle crash, it may help to verbalize awareness of certain safety issues: "It looks like there are three cars involved. I don't see spilled fuel or downed power lines."

• *Avoid sleepiness.* At times, the driver may feel sleepy while driving, especially on long transports. This is especially true on 24-hour shifts. There are various strategies that can help: First, do not rely on caffeine or sugar for energy There is caffeine in many colas, chocolate and over-the-counter medicines. Both sugar and caffeine cause a rebound drop in energy a few hours later, which can generate a vicious cycle as follow-up efforts to generate energy are made with yet more sugar and caffeine. Both may also prevent restful sleep when the chance finally arrives. A better alternative is fresh air, which boosts energy levels, as does ten minutes or so of deep breathing. Shake off drowsiness by opening the EMS unit's win-

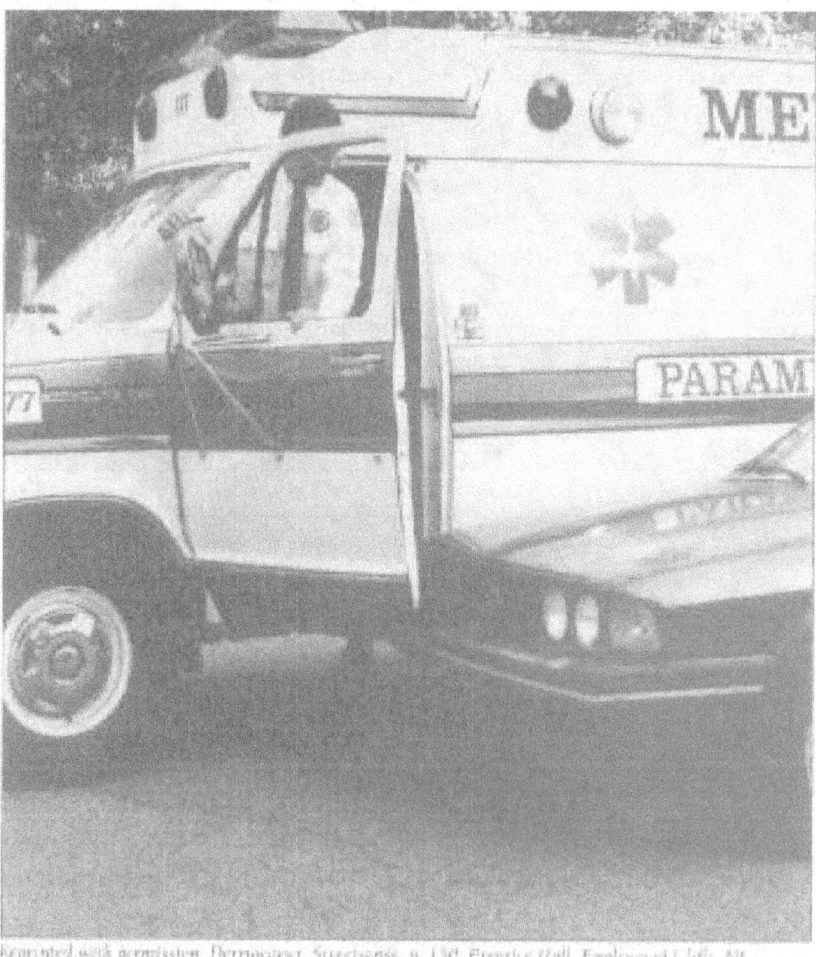

Reprinted with permission. Dernocoeur, Streetsense, p. 130, Prentice Hall, Englewood Cliffs, NJ

Figure 2-4: When opening emergency vehicle doors, be aware that on-coming traffic may not provide adequate clearance for a safe exit.

dow, or getting out of the EMS unit for a few minutes. Stretching also helps.

Some EMS personnel take prescribed medications that can cause sleepiness. If one's ability to perform the job is affected, the safe course is to admit the situation and avoid driving when the medication is interfering. If using antihistamines, use those that cause less drowsiness.

The Patient Compartment

The patient compartment is not typically thought of as being risky However, it is not a very safe place. Although it is as familiar as home to EMS personnel, serious and fatal injuries have occurred to prehospital providers who were in it. Certain strategies can promote safety while in the patient compartment:

* *Never let tunnel vision erase the need to protect yourself.* To allow patient care to totally consume your attention is to ignore the first rule of EMS: personal safety. You must remain consistently aware of both normal and unusual movement of the EMS unit, including changes in momentum and direction, speed and smoothness of the ride.

* *Use seat belts at all times when patient care allows.* Not using the seat belt in the patient compartment is a long-standing habit in EMS that will take considerable effort to alter. Prehospital personnel avoid using seat belts in the patient compartment because restraints are perceived to prevent ready access to the patient and equipment. However, evidence is increasing, both anecdotally and in the medical literature, that it is possible to provide competent patient care while wearing seat belts during transport. One group of researchers found that ALS personnel felt they needed to be unrestrained for providing patient care only an average of 41 percent of the time during the 148 patient encounters studied. In those providers' perception, the possibility of being securely belted existed more

Figure 2-5: Buckle up and clear the dash of all moveable objects before putting the emergency vehicle in motion.

than half of the time. Even in cardiac arrest situations, seat belts could be used up to 18 percent of the time.[7]

There are compelling reasons to wear seat belts or harness restraints in the patient compartment. Obviously, they can increase the chance of surviving a major crash. But sometimes even minor deceleration is enough to cause injury or even permanent disability, One paramedic was in the patient compartment, travelling without lights or siren, when the ambulance was involved in a low-speed (under 15 mph) crash. He was thrown against the door handle. The resulting paralysis left him with only minimal use of his arms. Others have also been hurt or killed.

• *Secure the patient.* Do whatever is necessary to keep the patient safely in place regardless of the ride. This safety issue is different from issues of restraining violent patients. It is not enough to think a patient will be safe because one or two straps are buckled. In sudden deceleration, patients have slipped out from underneath loose straps and become missiles-sometimes flying forward into the drivers compartment. It is also not enough to assume that patients immobilized for spinal precautions are secure. Without being adequately secured to the stretcher, the patient and the spine board can become missiles. Proper safety tie-down systems on the stretcher are the equivalent of using seat belts for people who are seated.

Systems for securing patients vary, There should be at least three belts, one at the chest, one over the iliac crests and one just above the knees. To avoid the chance that the patient might slip out of the stretcher's belts, some EMS organizations use an over-the-shoulder system, which prevents movement toward the head end of the stretcher. Such systems also provide good lateral support. A strategy to prevent the patient from slipping toward the foot end of the stretcher is to weave the chest belt under the arms in the axilla and then over the chest (see Figure 2-6).

Always remember to fasten the stretcher belts snugly and to refasten or tighten belts that are undone or loosened en route. Be sure no belt or tie is dragging; they can catch in the wheels and trip the stretcher when it is being rolled along.

Children have special needs during transport. Child restraint seats are required by law in most states. A good option is to use a child's own car seat; however, a child may not always be using a car seat. Similarly, some vehicles have built-in child seats that are not removable. Child restraint seats for use in EMS units are commercially available (see Figure 2-7).

• *Strap down equipment.* Anyone who has ever seen the back end of a stocked EMS unit after a rollover has a strong appreciation for the importance of tying down equipment. Loose, flying equipment has caused major and permanent injuries to EMS personnel during rollovers. EKG monitors, radios, jump kits and sharps boxes are not usually stored in a cabinet. Squad bench lids and sliding cabinet doors without latches may spill equipment including IV solutions, needles, splints, small oxygen bottles and neck immobilizers.

Examine the patient compartment with an eye for what equipment would not stay in place if the EMS unit were in a bad collision or rollover. Then find appropriate methods for strapping down that equipment. In order to meet the needs of EMS personnel at situations of great intensity, the latches should be the "quick-release" type using positive-latching devices.

• *Develop the habit of "hanging on"* Certain tasks require being off the squad bench or captains chair. Moving around the patient compartment is part of the inherent risk of EMS. For example, when CPR is performed in a moving ambulance, the person doing chest compressions must bend over the patient, arms straight, with appropriate positioning to push downward. The caregiver may be at the head one moment to check airway status and

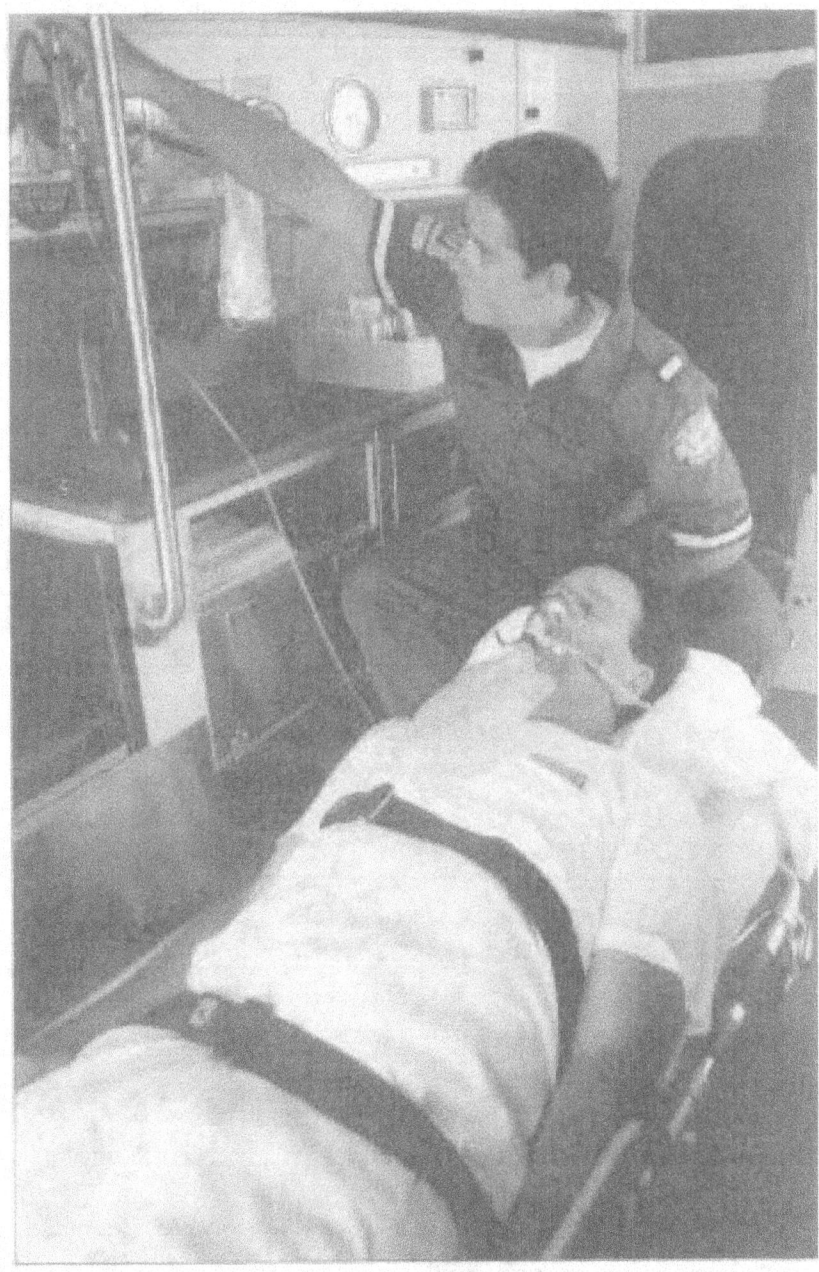

Figure 2-6: Patients should be strapped at the chest, hips, and knees, with chest straps running under the arms. Equipment should also be strapped securely. Whenever possible, the EMS provider should also wear a seat belt.

at the arm a moment later to monitor the IV.

Hanging on properly can minimize the chance of being thrown.[*] The process follows rock climbers' principles: the hands and feet constitute four points of attachment to a rock wall. To stay on the cliff requires use of at least two of these extremities at all times. Should a hand- or foothold fail, there is no backup. Rock climbers with an appreciation for safety hug the rock with at least three extremities at all times; only one extremity seeks a new hold at a time.

The principles are the same in the EMS unit, although an added point of attachment is the seat of the pants. Like the rock climber, the prudent EMS provider consistently uses three points of attachment. For example, when sitting on the squad bench, the EMS provider can use both hands if the feet are placed flat on the floor, spread apart for the best base of support. An extra advantage can be gained by hooking a toe under the stretcher bar (see Figure 2-8). Drop the center of gravity further by squatting between the stretcher and the squad bench when circumstances permit.

Walking or leaning across the patient compartment is especially hazardous because the center of gravity is higher than when seated. Hold the ceiling bars, moving arm-over-arm.

Hanging on takes on new meaning in the context of cardiac arrest. The

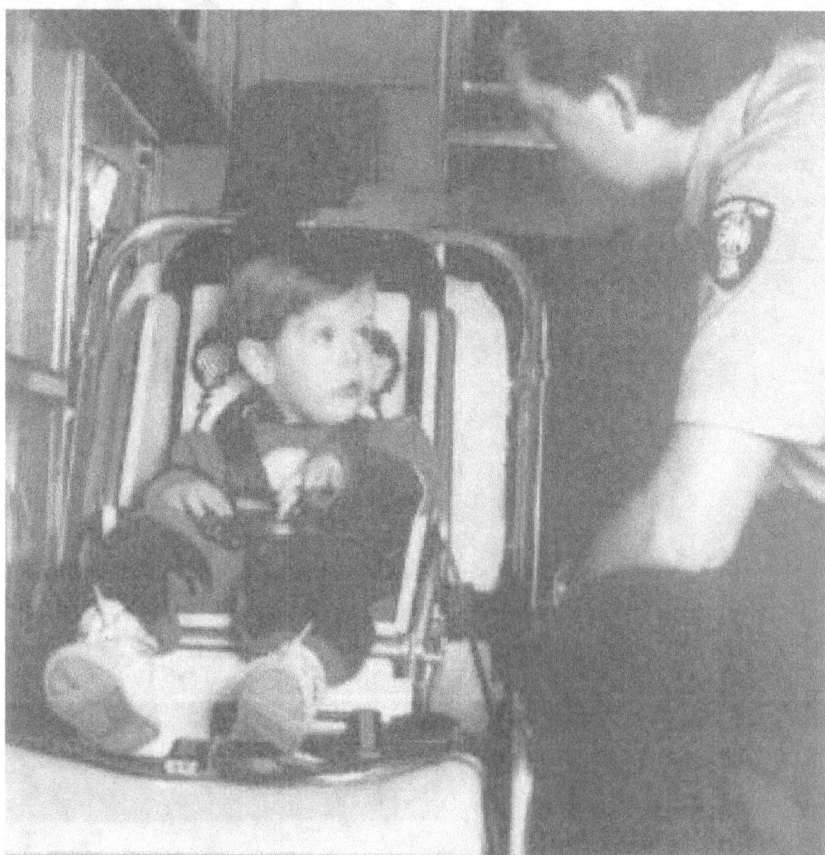

Figure 2-7: Small children can be transported in their own car seats, which can be strapped to the stretcher or in the captain's chair. Avoid strapping them on the squad bench, since rapid deceleration could cause lateral torque on the child. A child in a car seat who has sustained possible spinal injuries should be immobilized. Tape a horseshoe-shaped towel or soft pillow around the head to the shoulders.

person doing compressions has a high center of gravity because of the necessary positioning. For shorter people, a wide stance may be difficult. One remedy is to have a backup helper who is wearing a lap belt seated on the squad bench hold the compressor's belt. Optimally, the person holding on should be able to switch with the compressor. This technique will not help much in the event of a bad crash, but it can help minimize the risk of injury from loss of balance due to the changes in vehicle momentum that make every ambulance ride challenging. Another remedy is to use a safety harness.

One alternative is to use a mechanical chest compressor during trans-

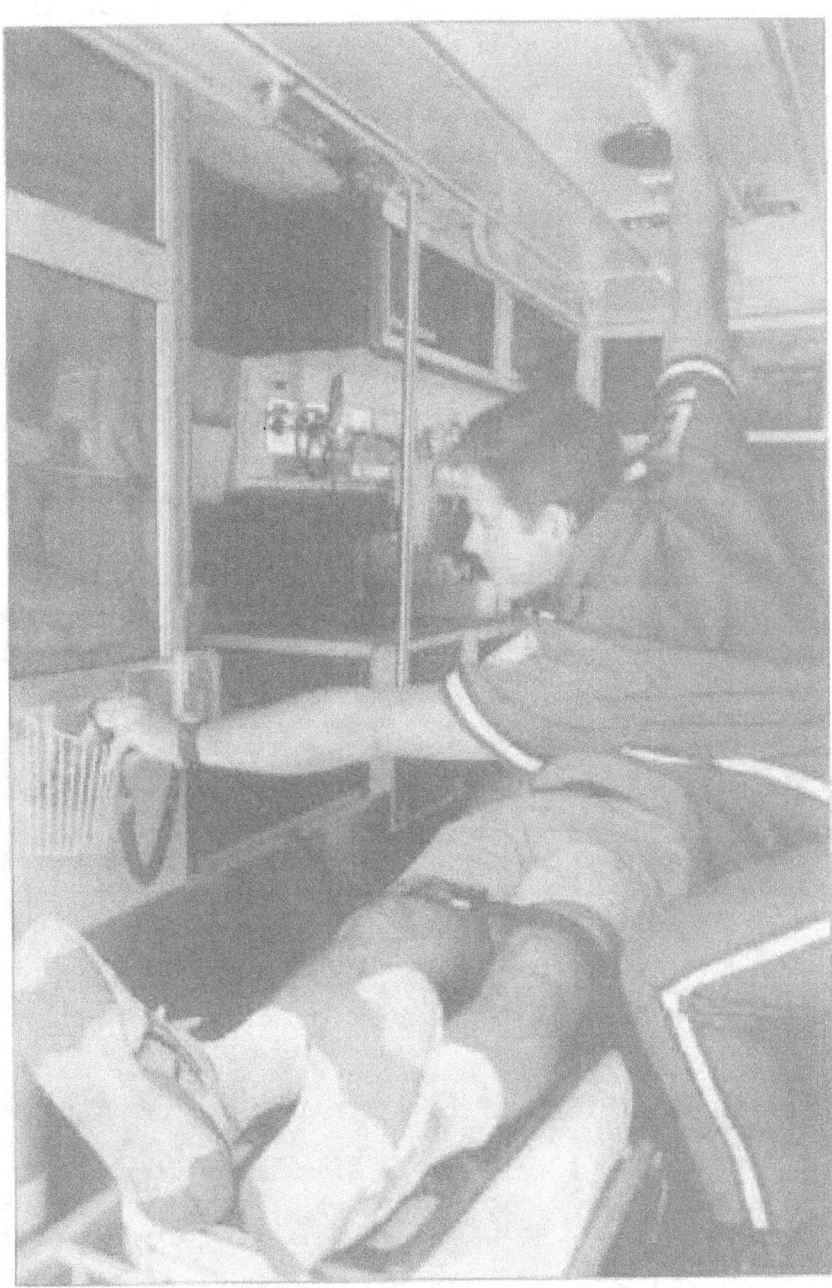

Figure 2-8: Develop the habit of always hanging on with one hand, especially when reaching for equipment. Keep feet flat on the floor and spread, to provide a broad base of support. The squad bench railing is a convenient handhold as well.

port. A research team compared CPR during ambulance transport, using manual CPR and two different mechanical methods. According to the researchers, "[a] gas-powered compressor provided adequate compressions during 97 percent of the tests, while manual compressions only resulted in a 37-percent adequate performance rate."[9]

• Develop the practice of "bracing." The habit of bracing is another strategy for safety in the patient compartment. It nicely complements the principle of hanging on[10]. This is the use of counter-pressure against different surfaces in the EMS unit to "wedge" the body into position. For example, when seated on the squad bench, counter-pressure by the lower legs against the squad bench and stretcher help wedge the EMS provider in place. This allows for steadier patient care and also minimizes the chance of being thrown if the EMS unit bumps, sways or stops short unexpectedly (see Figure: 2-9). Learning this technique requires some patience, but once ingrained, it is done unconsciously,

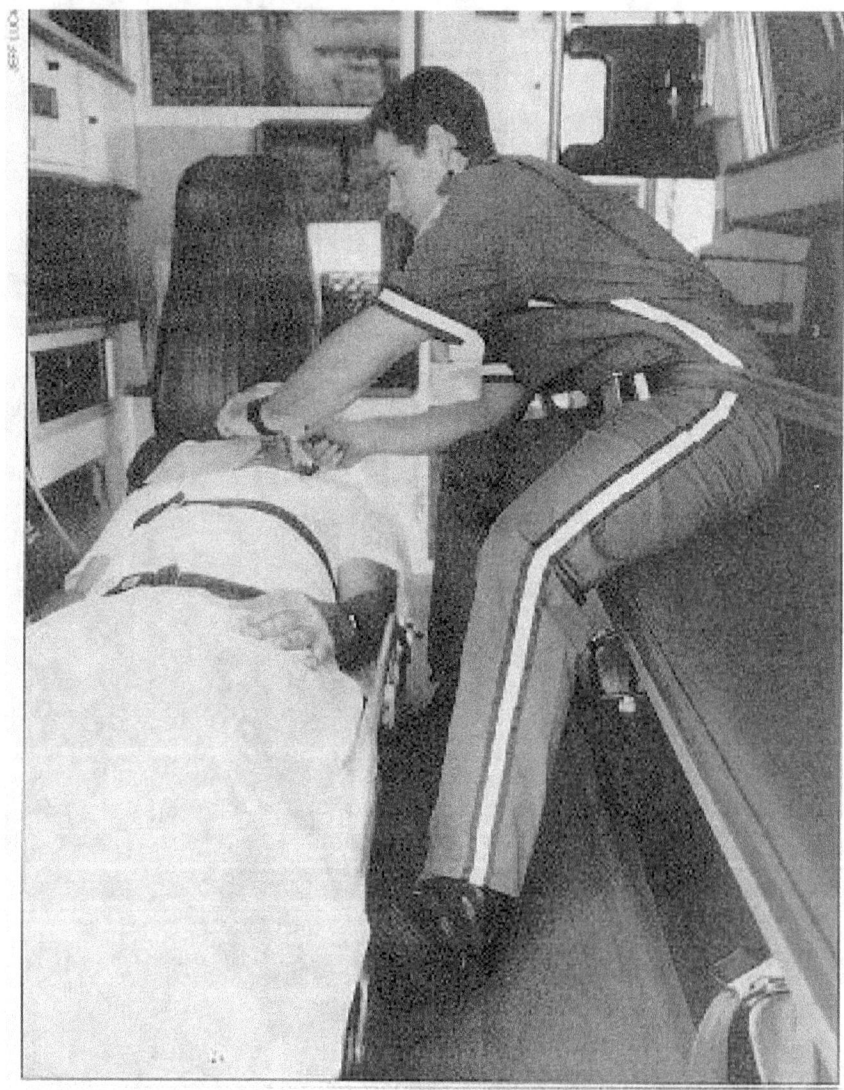

Figure 2-9: When both hands are busy, use counter-pressure to help brace against bumps, turns, and sudden changes in momentum. This EMS provider is demonstrating the counter-pressure that comes from hooking the toes under the stretcher railing.

The captain's chair (the seat at the patient's head) is a source of bracing, since the entire back of the chair acts as a cushion against sudden deceleration. This is better than the torque generated against anyone seat-belted onto the squad bench, which lies sideways to the line of travel. In particular, pregnant patients who will be sitting up may be more comfortable and safer using the captain's chair.

• *Be careful of hazardous equipment.* Some EMS organizations now mandate that the ambulance be stopped when the attendant is starting an IV or using other needles, and also during defibrillation. The dangers inherent with the use of this equipment merit consideration. Stopping the vehicle is the safest action to take. At the least, the patient attendant should tell the driver that a hazardous procedure is about to be performed so that the driver can drive particularly smoothly and steadily.

Principles of Safe Driving

EMS personnel are asked to drive vehicles that are usually much larger and heavier than their private vehicles, yet many have never had formal training about emergency medical vehicle driving and maintenance. Driving "strategies" are often learned through on-the-job practice and from others who may impart poor driving habits.

Anyone entrusted with driving an EMS vehicle must attend to the law of due regard. That is, despite the urgency associated with medical crises, the driver must perform driving tasks safely and with consideration (due regard) for the people both in the emergency vehicle and in surrounding traffic. Examples of exceeding the law of due regard might include using emergency warning equipment inappropriately, failing to yield right of way, speeding and following too closely. Emergency mode is a request, not a demand, for other traffic to yield; safe passage is a privilege, not a right. It is granted by the other motorists on the roadway. The operator of an emergency vehicle is not exempt from traffic laws and may be criminally or civilly liable if death, injury or property damage result from a crash.

Basic principles of safety related to driving an EMS unit are:

• *Driving with lights and siren can be dangerous.* Emergency mode is more dangerous than non-emergency mode primarily because citizens often react in unpredictable and unusual ways. Some do as the law requires by pulling to the right and stopping. Others pull to the left, stop dead ahead of the oncoming EMS unit or decide to race it. Avoid the trap of complacency. A crash can occur anytime-especially when a driver stops being attentive.

Certain stress-related factors about emergency lights and sirens are believed to contribute to poor performance and the potential for crashing. An incident that occurred in Virginia in 1989 involved the increased stress inherent in driving emergency equipment. In this case, a train travelling 77 mph struck a fire truck that was crossing the railroad track at a crossing marked by a sign but no barrier. In the report compiled after the incident, the National Transportation Safety Board concluded:

The driver's failure to determine that it was safe to proceed must be attributable not only to the usual pressures experienced by fire service personnel when responding to an emergency but also to the series of events that occurred during Wagon 7's response to the fire call. What should have been a routine response to a fire that initially posed little threat to life or other property became-in less than 6 minutes a response involving a succession of performance errors or omissions that resulted in a steadily increasing level of frustration and stress . . .

The driver, despite having his window rolled down, showed no sign of noticing the oncoming train as the emergency vehicle began to cross the tracks. The report continues:

" . . . excessive stress can lead to substandard performance. When a persons arousal level is unduly increased by stressors, the focus of attention is narrowed to performance of the task perceived to be the most important [in this case, arriving at the scene], while the quality of the performance of any peripheral task(s) deteriorates [in this case, driving the vehicle across the tracks safely] ."[11]

The term referred to in the report was "sirencide"—a phenomenon "used to describe the emotional reaction of emergency vehicle drivers when they begin to feel a sense of power and urgency that blocks out reason and prudence, leading to the reckless operation of the emergency vehicle."[12]

Because of the implication of urgency associated with using the lights and siren, stress builds up in the driver and passengers. Research shows that people driving EMS units "are affected by the continuous sound of a siren. Tests have shown that inexperienced ambulance drivers tend to increase their driving speeds from 10 to 15 miles per hour while continually sounding the siren. In some reported cases, drivers using a siren were unable to negotiate curves that they could pass through easily when not sounding the siren."[13]

• *Use emergency warning lights and sirens appropriately.* The practice of dispatching EMS vehicles in emergency mode is increasingly being questioned and balanced against related risks. Few communities will tolerate the tragedy caused by outmoded practices or excessive use of emergency mode. One estimate is that 40 percent of all emergency responses are made in emergency mode, yet half that amount are actual medical emergencies, while only 1 percent to 5 percent are life-threatening. EMS agencies with greater than 5 percent of returns to the hospital in emergency mode are likely to have an operational problem that could ultimately prove costly.[14] Emergency medical vehicle crashes cause enough loss of life and health, enough destruction of expensive equipment and enough erosion of public trust in EMS providers' mission to "do no harm." But when a pedestrian is hit and killed by an EMS unit responding to an ill-defined "man down" (especially when the call nature turns out to be minor), the repercussions can be particularly widespread. When an EMS unit uses emergency mode to transport a patient with a non-life-threatening condition and a crash happens, the backlash can be even more consequential.

One way to minimize use of emergency mode responses is by using the Emergency Medical Dispatch model (see Chapter 4). Research shows increasingly that agencies that use the EMD program properly can minimize the need for emergency mode and enjoy diminished crash rates-without negative consequences for patients.[15] A well-trained EMD knows how to gather relevant information to pass along to responders and can project a sense of control through tone of voice and choice of words. The calmer the EMD, the calmer responders tend to remain. However, if a dispatcher gets flustered or excited by the events at hand, vocal inflection reflects loss of control. Theoretically, the higher the pitch of the voice of the dispatcher, the faster the EMS unit may be driven by the responders.

When using emergency mode, do not use high beams; these can camouflage the emergency lights and confuse or blind oncoming motorists. To minimize confusion, headlight flashers should be used only in daylight.

The hand-held spotlights common to emergency vehicles are not part of the emergency warning system, and should be used for locating streets signs, house numbers and other landmarks for navigation. In rare cases of heavy traffic, the spotlight may be used in a rapid sideways flashing movement across the drivers' rear-view mirrors ahead as a last resort to gain attention. Spotlights should never be used in a manner that could blind another driver, even momentarily The driver's partner should use the spotlight only at the driver's express request. Used carelessly or inappropriately, spotlights can cause crashes for which the operator may be liable. Hand-held spotlights should be properly secured when not in use, preferably in such a way that they do not present a hazard in a crash. (Spotlight hooks also should be installed in a crash-safe location.)

• *Obey local standard operating procedures.* Jurisdictions have rules about the use of emergency mode. Most require lights and siren to be used during an emergency response; others allow some discretion to minimize use of emergency mode in the interest of community safety, When legally permissible, there are many situations in which arriving with less fanfare has merit. In these cases, turn off emergency flashers and siren early In situations of potential interpersonal violence, shut down one to two blocks away, and drive to the scene obeying all the rules of the road as a non-emergency vehicle. This prevents attracting a large crowd of curious spectators (some of whom may be hostile toward anyone in a uniform). When arriving at a scene where other rescue personnel are already at work-such as a traffic incident-shutting down warning equipment a block or two early may also help prevent the interruption of patient care in progress.

• *If using warning equipment, use all of it.* Some drivers, in trying to minimize patient stress and for personal safety, do not use the siren all or even part of the time when emergency lights are on. They do not want to upset people, such as the patient, the patient's family or neighbors. They wish to avoid drawing a large crowd to the scene. If local rules allow driver discretion, be consistent; either use all emergency equipment, or none. When not using emergency mode, drive with appropriate non-emergency technique. The mental shift in concentration is especially critical when making a transition from one mode to the other. When parking at the scene, remember that the EMS unit has not arrived safely until the gear shift is put into "Park" and the emergency brake is set.

The importance of following the rules may seem obvious, but the seriousness of doing so warrants emphasis. Operating in any way outside the scope of a standard, written policy places the driver at great legal risk if something goes wrong. If the local rules mandate a complete stop at each red traffic light, stop sign or railroad crossing, do it. This is especially crucial since most emergency vehicle crashes occur at intersections.[16,17]

• *Provide a smooth ride for passengers in the patient compartment.* Be acutely aware of the impact of the ride on those in the back of a moving EMS unit. Colleagues providing patient care may be unrestrained. Patients are already nervous. Minimize swaying by avoiding gutter lanes and abrupt lane changes. Avoid rapid deceleration and jolting stops; with correct technique and advance planning, it is possible to stop an EMS unit almost imperceptibly. If road hazards (such as railroad tracks or a suddenly red traffic light) occur, front-to-back communication can give those in the rear a chance to hang on tighter or brace better.

Become completely familiar with the stopping distance needed for each EMS unit. It is longer than on a personal vehicle. EMS vehicles fully loaded with equipment, crew and patients may exceed their maximum GVW To avoid rear-end collisions, maintain safe following distances. Leave one car length for each 10 mph the EMS vehicle is traveling between it and the next car on the road. A vehicle moving 40 mph should be four car lengths behind the next car. If estimating distance is difficult, allow 3 to 4 seconds between vehicles; when the car ahead passes a point of reference (such as a light pole), you should be able to count "one-one-thousand, two-one-thousand, three-one-thousand, four-one-thousand" before reaching the same point of reference.

In certain circumstances, double the following distance: when transporting a patient; at night; when fatigued; and in conditions of rain, fog or smoke; and on loose road surfaces such as dirt or gravel. In conditions of snow, packed snow or ice, triple the following distance.

• *Pay particular attention when crossing intersections.* More crashes occur in intersections than anywhere else. Visibility is often impaired by buildings or traffic, and it is easy to misjudge traffic situations. Motorists may not notice the approaching emergency vehicle. Other emergency vehicles may be approaching simultaneously from another direction. To minimize the chance of a crash, maintain safe following distances from other vehicles. Stop before the front bumper extends into the pedestrian lane or first white line. If another vehicle is in the lane, leave one vehicle length between the two vehicles, or use the guideline of leaving enough space that the rear tires and bumper of the car in front are still visible. Leave the brake foot in place until the car in front has already begun accelerating; anticipate that it may stop short at any moment. Before proceeding, look left, then right, then left again before entering the intersection.

- *Watch for multiple responses and colliding sirens.* No one is served when emergency service vehicles collide. Because it is typical for two or more emergency vehicles to respond to the same scene, the potential is great, especially at intersections. In just 90 days in 1979, there were four such collisions in one state, one involving the death of a 44-year-old firefighter. His engine company collided with the truck company from the same firehouse as the two vehicles approached the scene from different directions. Six other firefighters and a civilian were injured, a valuable pumper was destroyed, and the new tractor-drawn aerial sustained heavy damage. Equipment replacement and repair costs were over $140,000.

Another incident involved a brand-new tanker making its first emergency response, A third involved a heavy-duty rescue squad that was hit at an intersection. The driver was flung out the door, but he held onto the cab's open door as the squad careened down the street and ran into a parking lot, damaging numerous cars. The entire crew was hospitalized. (The call: a faulty automatic alarm.) In yet another incident, a medic unit stopped short when it missed a turn, and the ambulance following it plowed into it from behind.[18] All of these cases give lessons to those willing to heed them. (This is not a new problem: the March 1935 issue of Volunteer Firemen reported on an analysis of 75 typical cases and discovered collisions with other emergency vehicles to be the most common type of crash, resulting in 38 of the 100 fatalities.)

It is dangerous to be in a convoy of vehicles using emergency mode. If a justifiable reason for an emergency escort arises, follow no closer than 300 feet. Be especially vigilant at intersections; other motorists are likely to think there are no other emergency vehicles and may enter the intersection without due caution. Also, if the first vehicle is in a crash, there should be adequate time for a following driver to react appropriately.

- *Night driving. Driving at night is harder than driving in daytime,* for various reasons. Other drivers are more likely to have been drinking alcohol or using drugs at night. Landmarks are harder to pick out. Map reading is harder. Fatigue may be more profound. Animals or pedestrians in the roadway are more difficult to notice early. It is hard or impossible to establish eye contact with other drivers, It may help to dim the instrument panel lights and to avoid using high beams or front strobes in snow or fog conditions. Also, a relationship exists between smoking cigarettes and night vision which could be a safety hazard. The aftereffects of both primary and secondary smoke are believed to decrease night vision by up to 40 percent.[19]

- *Backing up.* Backing crashes are a very common cause of equipment damage, according to one EMS insurer, even though a majority of agencies have policies intended to prevent them. Although not usually a high-risk safety issue, backing incidents highlight the disregard found in many work settings toward maintaining appropriate emphasis on safety

Backing an emergency vehicle is different from backing a passenger car. The driver must use the side mirrors, since the view through the patient compartment is usually inadequate. EMS vehicles are often nosed into tight spots upon arrival (with good intentions of leaving carefully later).

Always post a spotter 8 to ten feet from the left rear bumper, in eye contact with the driver, using mutually understood hand signals. Some departments also require a right side spotter (see Figures 2-10 and 2-11) Establish voice and hand communication before beginning to back. Never hurry through a backing procedure.

* *Never leave a running vehicle unattended.* They have been known to

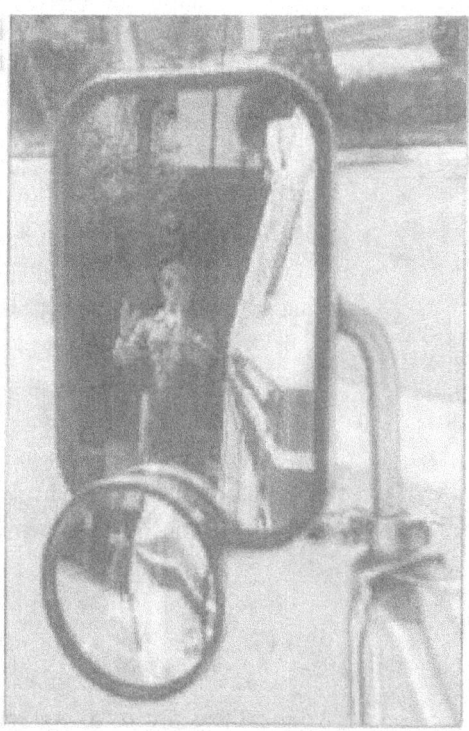

Figure 2-10: Always have a spotter visible in the mirrors when backing an emergency vehicle. Use mutually-understood hand signals. This person is signaling the distance remaining for backing up.

Figure 2-11 : Be sure the spotter stands well away from the backing vehicle. This spotter is signaling the driver to stop.

be stolen, even from in front of a patient's address. Even at the emergency facility, nothing is sacred; one paramedic was cleaning up the patient compartment when the unit was driven away-and not by his partner. In another instance, the driver left the unit briefly to summon his partner for a call; as they exited from the emergency department, they could see their vehicle was exiting the driveway.

Basic Vehicle Maintenance

Emergency vehicles should be inspected at the beginning of each shift. Vehicles not driven in shifts, such as with many volunteer systems, should be inspected daily, Even seldom-used backup vehicles should be inspected at least weekly; these vehicles are typically pressed into service during a mass-casualty situation, when equipment availability is most critical. Emergency vehicles are driven hard, even in the gentlest EMS systems; lives depend on their roadworthiness. Inspection should include:

- tires (tread wear and tire pressure)
- brakes
- warning lights and siren
- headlights and running lights are clean and working
- windshield wipers
- noting whether windows are clean inside and out
- mirrors
- all fluid levels
- ease of starting
- safety equipment (such as fire extinguishers)
- general cleanliness and overall condition of vehicle body

A written record of inspection for each vehicle should be routinely submitted to the appropriate person, who should read it and follow through on needed repairs. Actions needed to bring the vehicle to operating standards should be taken immediately.

Summary

Driving an EMS unit safely is a challenge. With increased emphasis on this aspect of emergency care, lives and millions of repair dollars can be saved. Attention to actions done in the patient compartment is particularly overdue for scrutiny and adjustment; the habits of hanging on, bracing and using seat belts and compartment closures will help. Improved driving habits, too, will result in improved safety for the EMS professionals and the public they serve.

References/Endnotes

1. Clark Hendrixson, "Nightmare on Elm Street," Rescue-EMS Magazine, September-October 1991, p. 44.
2. Gordon M. Sachs, "It Can Happen To You," Emergency Medical Services, August, 1991, p. 72.
3. Robert Elling, "Dispelling Myths on Ambulance Accidents,"JEMS 11:7, July 1989, pp. 60-64.

4. "Statistics on Ambulance Accidents," AzStar Casualty Company, LSCT-26, 2-11-92.

5. Fatal Accident Reporting System 1989, National Highway Traffic Safety Administration (Washington D.C.: US Dept. of Transportation, 1989), p. 2.

6. Elling, p. 60-64.

7. Cook, RT Jr., Meador, SA, et al. "Opportunity for Seatbelt Usage by ALS Providers," Prehospital and Disaster Medicine 6:4 (October-December 1991) pp. 469-484.

8. Kate Dernocoeur, Streetsense: Communication, Safety and Control, 2nd. Ed. (Englewood, NJ: Brady, 1990), pp. 179-183.

9. Edward R. Stapleton, "Comparing CPR During Ambulance Transport: Manual vs. Mechanical Methods,"JEMS, September, 1991, pp. 63-72.

10. Kate Dernocoeur, Streetsense: Communication, Safety and Control, 2nd. Ed. (Englewood, NJ:Brady, 1990), pp. 183-184.

11. National Transportation Safety Board report by Chairman James L. Kolstad dated January 4, 1991 regarding the incident in Catlett, Virginia on September 28, 1989.

12. Originally cited, with citation taken from the NTSB report referred to in the previous footnote, in Firefighter's News, August-September 1990, pp. 36-37.

13. H.D. Grant, R.H. Murray, Jr., and J.D. Bergeron, Emergency Care 4th edition (Englewood Cliffs, NJ: Brady, 1986), pp. 481-2.

14. "Statistics on Ambulance Accidents," AzStar Casualty Company, LSCT-26, dated 2/11/92.

15. Darrell Willis, "Dispatch Program" (in Letters to the Editor),JEMS, July 1988.

16. The relevance of Emergency Medical Dispatch to EMS safety is described in greater detail in Chapter 4: Principles of Teamwork.

17. Hendrixson, p. 45.

18. Elling, p. 60-64.

19. "A Tragic Loss-A Tragic Record," The Maryland Fire & Rescue Institute Bulletin.

20. "The EMS Street Survival Seminar," presented by Mike Taigman and Bruce Adams, Chicago, 1988.

CHAPTER 3
Lifting

Chapter Overview: The potential for back injury is very high in the emergency medical services because of the intense manual labor often required. In this chapter, the risk factors are examined (including the ergonomics of being in an emergency vehicle for hours at a stretch), proper lifting dynamics are described, and injury prevention tactics are offered.

EMS personnel must come to work prepared to do hard physical labor. Just arriving at a patient's side routinely involves carrying heavy bags or cases, suction units, EKG monitor-defibrillators and other heavy, bulky equipment. Many of those patients are then placed on backboards or stretchers (often a significant lifting challenge) and wheeled to the EMS unit. Adding to the weight of the stretcher may be an oxygen tank and other heavy medical equipment. Getting to the EMS unit may entail lifting the stretcher over sidewalk edges, up and down stairs and around tight corners. The patient is loaded into the EMS unit, and unloaded and moved again at the destination. With 30 percent of the population exceeding ideal body weight,' the resulting load is often significant.

It is no wonder that back strain heads the list of injuries sustained by EMS personnel. One group of researchers looked retrospectively at 254 injuries that occurred in 3½ years in a busy urban EMS system with a strict policy of job-related injury reporting. An overwhelming majority-36 percent-of injuries were back strains, all due to lifting. (The next most common injuries were nonspecific contusions at 7 percent, toxic effects of gases at 6 percent and ankle sprains at 5 percent.) Back injury rates were significantly higher both in women and in basic-level EMTs. Of 481 days for which compensation was paid, 375 (78 percent) were due to low back strain."

Risk Factors

There are various reasons why back injury is so common in EMS. The exertion of heavy lifting is usually done either sporadically or nearly continuously, depending on call volume; each practice is dangerous. Some personnel await calls posted in the EMS unit, a place not noted for fostering good posture. Calls arrive regardless of when an EMS provider last rested or ate. Bursts of exertion on tired, stiff, unfueled bodies create an easy setup for injury,

Mental attitude contributes to increased risk of back trouble. Prehospital personnel are often young, healthy individuals, with a powerful sense of invulnerability They believe strength and health will last forever. Few understand that without proper care and attention, they could end up hurt and unable to work in prehospital care. A sedentary lifestyle, chronically poor posture, stress, being overweight or obese, lack of flexibility, poor physical condition-each contributes to the risks of back injury (see Chapter 13).

Proper Lifting Dynamics

Every situation is unique. Applying the basic principles of lifting in each lifting situation will result in more consistently safe lifts; attention to this detail is a vital self-care habit. Proper lifting dynamics include:

• Think through the situation. Anticipate and troubleshoot tight corners, stairs and small elevators, and evaluate better alternative routes.

• Determine which partner will lead. Be sure everyone involved knows who is in charge.

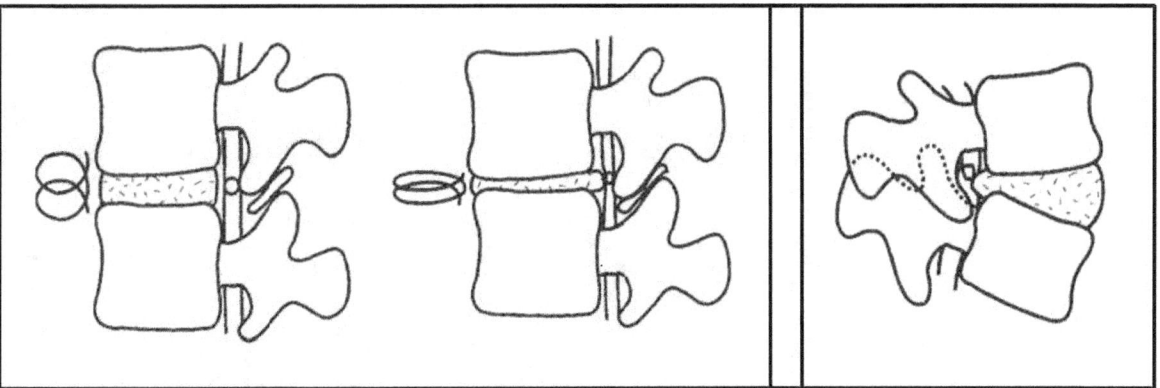

Figures 3-1 and 3-2 Anatomy of the spine and effects of weight lifting.

• *Communicate with helpers.* Get verbal feedback to be sure they understand the plan.

• *Check for adequate footing.* Many lifting situations are on difficult terrain. A slip at the wrong moment can spell disaster. The importance of wearing appropriate shoes cannot be overstressed (see Chapter 5).

• *Stay in the most balanced position possible.* As the load shifts, the leader should stop the action so that everyone can re-establish appropriate balance. For maximum spinal safety, weight-bearing force should travel through the vertebrae and discs (see Figure 3-1), not through the much-weaker and susceptible facet joints (see Figure 3-2).

• *Use the powerful leg muscles, not the back, to accomplish the lift.* Lift in a smooth, continuous motion, never with fast, jerking motions. Consciously notice whether the quadriceps muscles in the front part of the thighs are used every time. Proper positioning to use the muscles means bending the knees. This enables the back to remain straight, maintaining good spinal alignment. Bending over to lift is extremely dangerous to the spine, especially if one bends and twists simultaneously (see Figure 3-3).[3]

• *Exhale during exertion.* Holding the breath or grunting causes additional and easily avoidable strain. Exhaling forcefully but evenly during a lift can add power to the effort.

• *Keep the weight as close to the body as possible.* Avoid leaning or stretching to reach the load. Place hands palms up so the load is brought naturally toward the body when lifting.

• *Know personal limits.* Since lifting is a required skill in EMS, most lifting tasks should be routine and problem-free for professional EMS personnel. But every person is different. Abilities change. Self-awareness will help the EMS provider realize when personal exhaustion, minor illness or dwindling energy from a busy day causes certain lifting tasks to be risky. To overextend in order to avoid losing face is to jeopardize oneself, the patient and the others doing the lift. Never feel bad about asking for assistance to lift or move patients.

When deciding whether a particular lifting task is manageable, there are three considerations in addition to the weight of the load. First is the

Figure 3-3: Two people lifting, one with good lifting posture, the other without.

position of the load (for example, the 250-pound person who has slid into an awkward position underneath the steering wheel). Second is the distance and terrain the load is to be carried across (such as up a steep embankment). Finally, consider the repetition factor (meaning, is this the 15th heavy lift of the day? Or the first?)

Injury Prevention

Contrary to popular opinion, the back is powerful and resilient when properly maintained and nourished. Yet once a back is injured, it is three to five times more likely to experience subsequent injury.[*] Two significant back injuries will make re-injury predictable and likely. Injury prevention is the best approach for sustaining the ability to perform the important prehospital task of lifting.

The physical challenges of EMS may prevent some otherwise eager practitioners from being able to do the job safely, EMS organizations without a properly implemented back-health screening program may be asking for unnecessary expense if someone is hired whose physical condition precludes heavy labor. Medical examinations and back X-rays alone have not been shown to be effective screening tests. A screening program must test physical fitness and flexibility; to pass the Equal Employment Opportunity Commissions standards, it must be safe, reliable, practical and relevant to the job. Once employees are on the job, another preventative measure is periodic physical strength and agility testing. Personnel should expect to have to achieve and maintain a certain standard of conditioning.

"Back school" is another form of injury prevention. Realizing that personnel do not have the skills or information they need to prevent injury, employers are now arranging employee education on principles of safe lifting. Some back schools are post-injury facilities with education and training to help the back pain sufferer return to normal activity as soon as possible. Others are purely preventative. "Preventative back schools concentrate on the work site with a program content that is job-specific. They generally require one-half the number of hours that post-injury back schools require."[5]

Back schools are proving to be efficient and cost-effective. One research team studied eight industrial back schools that had operated for more than one year with programs customized for specific industries. The spokeswoman said, "The average reduction in back injuries was 25 percent, and the average reduction in lost work days was 50 percent."[6] Claims for back injuries for one company dropped from $215,000 to $48,000 in the first year the back school was begun-and to $13,000 in the second year.[7]

Elements of the lifting process that EMS personnel can control are the same elements which, uncontrolled, most often result in injury:

• *Avoid hurrying.* Emergency medical service work is often intense. The sense of urgency when another's life is in one's own hands is often a trigger for rushing. This diminishes the chances of carrying out a well-thought-out and coordinated plan.

• *Communicate clearly and constantly.* To assume others know how to go about a task is to court disaster. Even with a regular partner, it is best to put words to the plan each time. Otherwise, someone may not be ready to proceed. As a general principle, given the importance of the patients head and cervical spine, the person at the head is in charge of giving the commands for lifting and moving.

• *Avoid awkward positioning.* It is especially dangerous to the structures of the back for EMS personnel to lift from an awkward position. Yet patients seem to end up in the most interesting, inaccessible situations: the obese person unconscious and wedged into the bathtub; the driver in an upside-down, collapsed vehicle; the roofer experiencing extreme dizziness on a church roof with a slope of 65 degrees. Each is a mechanical nightmare for the vertebrae, spinal muscles, ligaments and tendons of rescuer backs. However, with foresight and effort, lifting dynamics can usually be improved. The basic principle is to find a way to effectively use leverage with the skeletal system rather than "muscling" through a lifting task, and to avoid twisting as well. Whenever possible, call for additional assistance.

• Think about optimal lifting dynamics before (and during) each lift Be conscious of the back and the various elements inherent to an optimal lift every time. The EMS provider has many things to attend to during an emergency call; lifting, being an event many consider peripheral to the greater priorities of the patient's needs, must become viewed as equally important. Make sensible lifting a habit. A red flag of impending back trouble is backache, especially at the end of the shift. Avoid it at all costs. If backache occurs, heed the body's natural signal that real trouble is brewing and seek appropriate assistance.

• Consider using a buck support belt. There are different back support devices that are intended to help reduce back injury, and there remain differences of opinion about how effective back support belts actually are. Back support belts should be chosen carefully and with the assistance of a person knowledgeable about their pros and cons. Use of such a belt does not replace the need for ongoing physical strength and flexibility training, Perhaps the best attribute of a back support belt, even above providing lower back support, is that its mere presence serves as subtle but constant reminder to lift properly.[8]

• Use help wisely. Although not standard procedure, there are rare circumstances when additional help from willing bystanders is used as a last resort. The EMS professional must use appropriate caution; allowing others to help can upset the load unless procedures are clearly communicated, Explain lifting commands. Do a "practice" run if possible; the cost in terms of seconds will be well repaid when the actual lift goes well. Remember that on-scene bystander/helpers are usually inexperienced, sometimes overwrought people. They may be willing, but they are untrained. And some may have underlying back dysfunction, which they momentarily forget or ignore in the excitement. Always inquire about pre-existing back trouble to prevent subsequent hassles in the event the bystander is injured.

Injury prevention requires ongoing attention to each element of the physical labor involved in prehospital care. EMS is admittedly risky, but caregivers can manage many of the controllable risks through physical fitness and health maintenance. All the elements matter: building muscular strength and flexibility, good nutrition, adequate rest, stress management and general self-care. These long-term wellness issues, which relate directly to this discussion on prevention, are discussed in Chapters 12 and 13.

Ergonomics

The study of the relationship between people and their workplace is known as ergonomics. The ergonomic factors of prehospital care are highly variable, given the unique character of every scene. The prehospital-

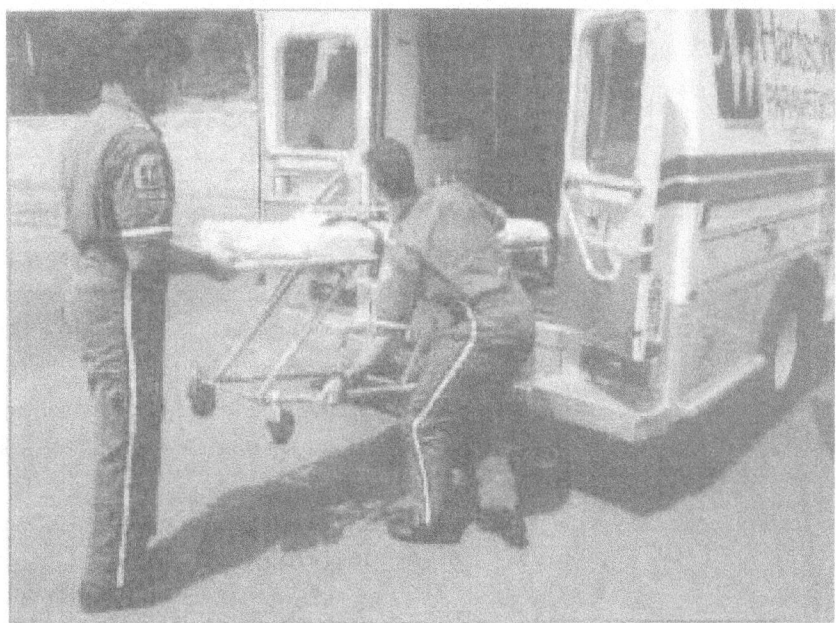

Figure 3-4: Modern stretcher designs help minimize the exertion needed to lift stretchers into the EMS unit.

providers "office" may range from being posted in an EMS unit all day to climbing trees and ladders, struggling to move obese patients onto hospital beds and leaning over chain-link fences to reach inside crashed vehicles. However, there are some things that can be done to improve the ergonomics of EMS work. These include:

* *EMS unit seats*. Office workers know the value of a proper work chair. Those who routinely sit in an EMS unit between calls deserve the same. Properly fitted lumbar supports are important.

* *Minimize driving*. People who spend more than half their working hours driving are said to be three times more likely to develop back problems than the average worker,[9] due to prolonged sitting that is followed immediately by physical exertion. Evidence suggests that the vibration of movement over roadways causes accelerated muscle and disk fatigue, which can result in degeneration of back strength."[10] Worth scrutiny is the increasingly common practice of dynamic dispersal of EMS units, in which crews rove among different districts of the service area; unnecessary shifting of crews from one place to another may contribute to the deterioration of back health among field personnel. One preventative strategy is to do frequent pelvic tilts; tighten the gluteal and abdominal muscles to flatten the lower back against the seat (or better, the floor, if a place to lie down can be found).

* *Proper use of stretchers*. There have been many improvements in stretcher design; it is no longer necessary to bend over to lower the loaded stretcher, then pick it up and step sideways to put it into the patient compartment. Roll-in stretchers allow the front end to be lifted onto the floor of the patient compartment while the wheels bear the load. The wheels are then lifted to the stretcher, and the unit is rolled inside (see Figure 3-4). In addition, mishandling the stretcher can cause metal fatigue. Equipment

that fails at a crucial moment, such as when changing the level of the stretcher, may suddenly jolt the spinal column and cause injury.

Summary

Lifting is one of the most recurrent tasks an EMS provider faces. Many do it for years, on thousands of EMS calls, without a problem; for others, back injury becomes a limiting factor in their ability to pursue a calling they enjoy. Good strategies for minimizing the chances of back injury have been described in this chapter (see Chapter 13 for more on physical fitness). Backs are not inherently weak. They can be strengthened. Back pain is not inevitable, provided a person understands how to avoid it and then pursues an appropriate prevention program.

References/Endnotes

1. Kate Democoeur, "Big Bodies: How to deal with them,"JEMS, September 1986, p.46.
2. Paul T. Hogya and Lloyd Ellis, "Evaluation of the Injury Profile of Personnel in a Busy Urban EMS System," American Journal of Emergency Medicine 8;4: 308-311.
3. Casey Terribilini, "The Spinal Column: Nifty Ways to Love Your Lumbar," JEMS, August 1992.
4. Casey Terribilini and Kate Dernocoeur, "Save Your Back: Injury prevention for EMS providers,"JEMS, October 1989, pp.34-41.
5. Terribilini, Dernocoeur, "Save Your Back," p.37.
6. Terribilini, Dernocoeur, "Save Your Back," p.37.
7. Fitzer, S, "Chelsea back program: one year later," Occupational Health and Safety, July, 1983.
8. Timothy L. Stokes, "Back Injury Reduction Project," for City of Chicago (Illinois) Fire Department, 1987, p.41.
9. M. Pope, J. Frymoyer, G Andersson, Occupational Low Back Pain (NYC: Praeger, 1984).
10. Gunnar Andersson, Don Chaffin, and Malcolm Pope, "Occupational Biomechanics of the Lumbar Spine," Occupational Low Back Pain (NYC: Praeger, 1984), pp.58-64.

CHAPTER 4
Teamwork and Attitude

Chapter Overview: A subtle but important element of safety rests with certain intangibles, including teamwork, leadership, public trust in the prehospital EMS system, effective interpersonal communication and avoidance of certain attitudes that can interfere with one's safety. Although hard to measure, many of these elements are cited when outside viewers visit prehospital EMS systems that have them. The element of teamwork includes properly trained emergency medical dispatchers; as the public's first link with the system, an EMD has power to establish a helpful, caring tone from the outset. A description of this aspect of EMS and how it can influence the safety of field providers is included in this chapter.

While teamwork is a fundamental element of prehospital care, it is seldom discussed. The need for partners to work together is obvious; the ability to do so may be challenging. Teamwork as a safety issue occurs on multiple levels, such as watching for physical hazards, coordinating lifting tasks, defusing intense emotions in others and negotiating traffic safely Even recognizing long-term situations that can interfere with safety (such as stress buildup or attitude changes) can be part of the team approach.

Good teamwork has powerful implications for the safety of EMS crews, both on a day-to-day level and on a broader scale. It ripples outward from individual EMS teams through one's entire EMS organization, through the emergency organizations in each community, and to the state and national level.

The structure of prehospital EMS response varies. Sometimes there are three or even four crew members per EMS unit, especially in fire service first response and volunteer systems. In such larger groups, a clear chain of command usually helps establish on-scene leadership. The team functions with traditional hierarchy of rank and order. But for thousands of other EMS workers, a team consists of two people-one to drive and the other to attend to the patient. In some EMS systems, partnerships are formed and maintained for months or even years. Two partners who share enough EMS experiences often develop an extremely special bond. Teamwork in such partnerships can develop to the point where calls are handled almost without speaking, so familiar is each with the others ways,

In other cases, individual EMS personnel have to work with a variety of partners, sometimes nearly every day. Not knowing one's partner well increases one's vulnerability due to a limited opportunity to develop teamwork; neither person knows the others style or prehospital care philosophy. For example, two paramedics in a system without regular partners responded to a notoriously rough part of town, and encountered two men trying to restrain their brother, who was high on the mood-altering drug PCP. One paramedic put a headlock on the patient. While doing so, he was backed into the corner of the kitchen and had no way out. His maneuvering apparently seemed too rough to the other brothers, who started advancing threateningly. The other paramedic, who could see that he and his partner were poorly positioned and outnumbered, was calling for assistance when the wrestling paramedic somehow got free. Luckily, they were able to run to the relative safety of the EMS unit.

One reason this call evolved the way it did was an uncoordinated effort due to an absence of teamwork. These paramedics did not know one another. Like most people, EMS personnel have "fight or flight" responses when threatened. Although professionals always seek to avoid a fight, the prehospital environment can be rough. Ideally, EMS personnel should never encounter hostility without accompanying law enforcement officers; in reality, it does happen. If a confrontation suddenly arises, some people fight and others flee. This pair was lucky to get out without injury

No matter how many rules an organization writes, the sheer variety of EMS situations demands a great deal of good judgment. It helps to know what one's partners natural inclinations (fight or flight?) are likely to be. Individual preferences and methods for approaching prehospital calls should be discussed whenever one works with an unfamiliar partner or group.

Public Trust and Teamwork

When emergency personnel at a scene are from various agencies or organizations, the need for solidarity and teamwork is just as important. In the unsophisticated perception of the public, "911" has arrived. Few lay persons distinguish between uniforms, logos or vehicles. All that matters is that help has arrived. The public trusts emergency services workers to uphold certain values and traditions. These lend a feeling of security within the community.

Problems can occur when even one emergency provider behaves in an unprofessional manner. Right or wrong, all the emergency service workers in the community may eventually be judged by those behaviors. What the public recognizes as "good care" has more to do with the level of emotional support and compassion displayed by the EMS personnel than the quality of technical medical care. Yet only one person in about 50 will file a formal complaint with an organization; others with a story usually just tell their friends and neighbors. A ripple effect of gossip may cause unprofessional emergency services to become viewed as adversaries, Emergency personnel may find little on-scene support from bystanders because they are no longer perceived as "good guys." Attacks against EMS personnel can-and do-occur more regularly in such communities.

The best defense against deterioration of public trust is to not allow it to begin in the first place. Building positive community attitude requires time and effort, through good teamwork and a unified commitment. If public trust is shattered for some reason and rebuilding becomes necessary, the effort is the responsibility of every member of the emergency services team. Every EMS crew must be trained and trusted to behave in ways that will consistently reinforce public trust. This is called good customer service.

Leadership

Leadership is far different from management. Like teamwork, leadership occurs on various levels. On a two-person crew, one is the leader. At the top of an organizations hierarchy is the "boss." Scholars have been unable to arrive at a consensus of the traits that make a good leader. Although leadership may be hard to define, good leadership is easy to recognize. Leadership goes beyond day-to-day handling of the many details of managing an EMS service.

In the intense arena of the prehospital world, good leadership is essential. A good leader helps EMS crews remember, as they head out into the community in their small, unsupervised groups, that everyone shares a common mission. Good central leadership can motivate individual crews to adhere to protocols and standard operating procedures. The idea of following all the standards of proper emergency medical care is instilled to the point where follow-through is reliable.

Effective formal leadership is sometimes absent in an organization. In such situations, leadership emerges-but the quality and authority of this emergent leadership is not necessarily good. Emergent leaders tend to be negative and resistant to authority. The traits that create good prehospital personnel-such as being control-oriented, action-oriented and individualistic-can also create a leadership nightmare when anarchy prevails.

From the perspective of teamwork, it is impossible to have good leadership without good "followership." EMS personnel are notably opinionated and are comfortable working autonomously. Because of these and other common traits, being a good follower is not easy for many Yet a good EMS leader can foster good followership. It helps to frequently remind everyone in the organization that the opportunities to lead do not depend on anything more than being in touch with a very basic concept, now used as an advertisement: "People don't want to be managed. They want to be led. Whoever heard of a world manager? World leader, yes. Educational leader. Political leader. Religious leader. Scout leader. Community leader. Labor leader. Business leader. They lead. They don't manage. The carrot always wins over the stick. Ask your horse. You can lead a horse to water, but you can't manage him to drink. If you want to manage somebody, manage yourself. Do that well and you'll be ready to stop managing. And start leading.'"

Communication

Good interpersonal communication is a vital aspect of good, safe teamwork in the uncontrolled world of prehospital care. It is the basis of having a partner say "clear right" to assist the driver across intersections. It is evidenced by a prearranged lifting message, such as "one, two, three, lift," with good eye contact being used during the lift, It is the mutually arranged signal that means "Get out! This is an unsafe situation." Understanding the needs of a partner is as important as understanding the needs of the patient or the bystanders.

With good teamwork, a team of two has four eyes watching for hazards, and the chances of noticing trouble are doubled. With good communication, various dangers are effectively relayed so that harm may be avoided by both partners.

Teamwork depends on clear and consistent communication because there are so many ways to approach emergency calls, even among long-term members of the same EMS squad. One new paramedic partnership had not discussed this. When a quiet, uncommunicative patient erupted into a fury that included throwing an empty water heater at the medics, one paramedic ran, while the other stayed to try to restrain the patient. In another case, two paramedics were fired upon while ascending a stairway. They fled-one to the left, the other to the right at the bottom of the stairs. Separated, they spent several fretful minutes worrying about each other until they could regroup. Issues of proper procedure aside, these cases demonstrate a lack of communication,

Although the need for good communication between partners may seem obvious, it is often neglected. Worse yet is communication between emergency agencies that respond together to the same scenes to provide different services. First responders, medical crews, extrication, hazard containment (either fire-related or law enforcement)-each deserves mutual respect. Taking time to learn others' names and to say hello can have a positive impact. Such an effort is like an insurance policy; on another call, an emergency services colleague from another organization may point out a safety hazard because of personal interest and mutual respect.

EMD: The Newest Member of the EMS Team

The communications center (dispatch) has various significant, if sometimes subtle, influences on teamwork and safety. The two main components of the Emergency Medical Dispatch process that help do this are:

* *prioritization of calls,* and
* *post-dispatch first aid instructions* (emergency medical care that bystanders can provide before EMS arrives).

Prioritization of calls means that the EMD, after a specific course of instruction and practice, can usually determine which resources will be most helpful at each prehospital scene. In outmoded dispatching systems, when a hysterical caller yelled, "Come quick! She's bleeding!" every available EMS provider was typically dispatched. Now, the EMD asks more questions. The result is more and better information, such as which part of the body is bleeding, whether the bleeding is arterial and certain other pertinent information. Often, callers with cut fingers panic and hold them under the faucet. They think they are bleeding to death. The educated EMD knows better and can safely send more appropriate resources-in this case, a single BLS unit.

In addition to questions about the patient's condition, the EMD often inquires about issues involving responder safety Has the assailant left the scene? Is the fire still burning? What weapons are involved? Improved call nature information and on-scene hazard identification provide an obvious safety benefit. Although there will always be some callers who cannot or will not provide such additional information, statistics show that more than 50 percent are able and willing.[2]

Using a rigorously tested system, call prioritization also allows EMDs to decide whether to send EMS personnel in the emergency mode. Minimizing the use of lights and sirens enhances the safety of all motorists. (Another subtle safety benefit of this practice is reduced wear and tear on emergency vehicles.) For example, in the scenario described above, the EMS unit could travel non-emergency to the scene, assuming the EMD uses the second major component of EMD: post-dispatch instructions.

Post-dispatch instructions are basic first aid recommendations given via telephone to people at the scene after emergency aid has been dispatched. EMD professionals know telecommunications techniques that can settle callers down, enabling them to provide aid effectively to the patient. Post-dispatch instructions might be as simple as having the caller with the cut finger remove it from under the faucet and apply direct pressure with a clean towel, or as complicated as giving CPR instructions over the telephone. There is almost no delay in assisting the patient-a response time much better than the best EMS unit could ever provide.

When an EMD can work positively and professionally with the caller, EMS personnel usually arrive to find a calmer, thus safer, atmosphere. This is a big improvement over the chaos that can occur when a crisis infuses normally nice people with adrenaline and intense emotions. Lay persons who like their public service providers are less likely to lash out, both on individual calls and generally When a documented save through telephone CPR instructions by an EMD is reported in the local media, public trust in the system builds. Everyone benefits.

By acknowledging the EMD as a member of the EMS team, the traditional "us versus them" relationship with dispatchers diminishes. The safety benefit to field personnel is that a collegial relationship encourages an EMD to mention hazards peculiar to the day (such as a day-long road closure) to oncoming field personnel.

The benefits of EMD just mentioned also help reduce stress. Less stress improves job satisfaction, which can help minimize turnover. This is a subtle safety factor. It takes time and experience to thoroughly understand the prehospital care environment so as to avoid inadvertent safety violations. A personnel roster that includes people with positive mental attitudes combined with lengthy experience enriches the EMS organization. Greater job satisfaction should lead to an interest in helping the organization succeed in its mission of good, safe prehospital care.

Attitude and Safety

A suitable mental attitude in EMS is paramount, but it is sometimes lacking. The impact of each rescuer's attitude is powerful. A compassionate, gentle, giving approach using appropriate choice of words, facial expression and body language "says" the EMS provider is truly interested in helping. This builds a positive relationship with the people being served. Yet people in EMS can sometimes appear arrogant and lacking in compassion. Or they may seem unnecessarily gruff, impatient or irritable. Nonverbal signals sometimes indicate that the EMS provider feels superior to those who called for help, or is disgusted by their circumstances. Negative or judgmental attitudes have an impact on rescuer safety because they can yield negative responses among the public.

EMS personnel should choose to avoid negative mental attitudes for many reasons, including personal safety. Constructive interpersonal dynamics help bring a positive, safer conclusion to EMS situations. In this section, negative attitudes are analyzed and positive solutions suggested:

Assumptions. EMS personnel often make inappropriate assumptions about the people or places where emergencies occur, as evidenced by the comments one often hears while riding in an emergency vehicle. Or they assume that structures will not collapse, nice dogs will continue to be nice, or that nothing else can go wrong, rather than ensuring that it won't.

This is dangerous. Avoid assuming that the "nice" part of town is safe, or that the "bad" part of town is always unsafe. One prehospital team, thinking they were responding to a person with chest pain in the fancy country club district, instead encountered a domestic dispute clouded by alcohol. As they entered, they were fired upon at close range. Fortunately, no one was hurt. Although advance information from trained emergency medical dispatchers is not infallible, it tends to be more accurate and thorough than in EMS systems without them. Had this call occurred after EMD capability began, it might have been reported differently

Preoccupation. Whether career or volunteer, EMS personnel must be prepared to drop everything, mentally and physically, to respond on emergency calls. Preoccupation typically stems from being absorbed in activities such as working on personal problems, studying or watching a TV program when something of greater priority (such as emergency notification tones) intervenes.

Each new scene is potentially dangerous for many different reasons. EMS responders preoccupied by anything other than the situation at hand have little opportunity to assess it properly to prevent getting too deeply into dangerous circumstances. One must have the ability to focus completely on the task at hand, and to be able to switch to other topics of focus quickly With practice, this is an achievable skill.

A remedy for preoccupation begins with the commitment to concentrate on information about each call as it becomes known while remaining flexible about alternative information that may arise. One way is the "clean slate" method. The practitioner creates a mental image of a blackboard or slate, which initially is totally clear. Distracting preoccupations have been set aside. As information becomes known, it is mentally noted on the slate, or updated, to avoid perpetuating inaccurate earlier perceptions. All the logistics of the scene, such as crowd and hazard control, patient care and other details, are stored in the same place. This prevents the jumble and confusion of a busy day from being carried mentally to the next call.

Complacency. Complacency is a feeling of security or safety in the unacknowledged face of potential danger. Compounding the situation is the idealism and invincibility of youth typical to some EMS personnel. Even though the majority of EMS situations are not threatening, falling into an attitude of complacency is hazardous.

Complacency may be more evident among rural EMS personnel, whose service areas tend to be friendlier. They often know the people they are serving, and the community supports the efforts of the local EMTs. EMS crews are generally greeted by people who are relieved and grateful for assistance. However, thousands of EMS runs with good outcomes can be wiped out by just one where a complacent EMS provider assumes that just because "nothing ever happens around here," nothing ever will.

Remedies for complacency involve mental readjustments similar to the effect of changing the channel on a television. To change the program, change the channel. Someone who recognizes a problem of complacency can select a different mental attitude. Two suggestions, which might be instilled through training or continuing education, are:

• *Never let crisis situations be a surprise.* Without becoming paranoid or negative, expect emergency situations to generate challenging logistics demands and difficult behaviors in others. Even on "simple" interfacility transfers, avoid complacency by acknowledging how disrupting the trip may be for the patient.

• *View each scene separately.* Do not allow the stream of calls already handled on a shift or in the past few weeks prescribe or bias an opinion about a new situation. For example, just because the last 40 shootings turned out to be false calls or no-transport situations does not mean this one won't need every ounce of effort.

Paranoia. After some time in prehospital care, it is possible to develop an attitude of paranoia. The prehospital worker who begins each new encounter with a "these people are out to get me" attitude is distracted by an unfair assumption. There is a difference between being paranoid and being careful.

Prejudice. Everyone has prejudices. Knowing this, the professional prehospital worker must set aside personal prejudices when on duty. That way, medical intervention and rapport-building remain appropriate. Most people in EMS conduct themselves above the levels prescribed as minimum standards by their agencies. This is due to the pride of doing a difficult and important job. However, there are times when each EMS provider tries even harder than usual, such as when caring for children or for co-workers. This demonstrates a spectrum of care. Prejudice sometimes causes that care to fall below minimum standards. This must be avoided.

Tombstone Courage. The phrase "tombstone courage"[3] vividly describes the potential results of careless heroic efforts. For example, one paramedic crew responded to a shooting at a bar. They were nearby, so they arrived quickly. Excited patrons waved them in. Because of the commotion (a primary reason to await backup), they elected to enter the bar. They were lucky; there was no need to buy them tombstones afterwards. But there was no insurance going into the bar that the end result would not be tragic. The remedy for tombstone courage is to learn to recognize personal limits and to say "wait" when every other impulse says "go!"

Rushing. The intensity of prehospital care sometimes causes people to lose sight of the importance of deliberate, cautious behavior. They begin to rush. They become careless. Rushing can result in serious harm. For example, one EMS crew was videotaped rushing off a scene with a patient in cardiac arrest. As they wheeled the ambulance cot along, a dangling belt got hung up under a wheel, tripping the cot and tipping it over. The cause: carelessness due to rushing. Checking to ensure that belts were fastened would have made all the difference. (Fortunately in this case, the EMS personnel sustained no back injuries while trying to catch the toppling stretcher-and the patient was not injured either-but it has happened to others.)

Because some of the negative mental attitudes just described may be mistaken for police-type tactics, EMS personnel who display them may find themselves victims of two negative consequences:

- they may be injured by people who would not have otherwise become hostile and aggressive, and

- they, along with all local EMS personnel, pay the price for a negative public perception. As mentioned, just one EMS provider with a bad attitude can spoil public trust toward the entire EMS team, especially if the event is reported in the media. For example, in Dallas in 1984, a call-screening nurse argued with a caller and failed to send an ambulance for the callers incoherent mother. His mother died. The situation was reported on the national news. What few people know is that the paramedics in Dallas were the ones who took the brunt of the publics outrage. Although the problem occurred at the dispatch center, the visible representatives of the fire department heard complaints, sarcasm and taunts. A hotline for the public to voice complaints about EMS in that city went from handling 8 calls a month to being completely overwhelmed.[4] Public trust is precious; once shattered, it cannot be rebuilt overnight.

It is said that when something unusual happens in a person's life, this person tells between 9 and 16 other people. News about how EMS crews handle their patients-good and bad-ripples out. Positive and negative public perception of the quality of local emergency medical service can be perpetuated just through word of mouth.

Summary

Building a good team, including those in the relative obscurity of the communications center, is an investment more EMS organizations should pursue-for reasons that include safety The skills needed to build a strong team require training and practice. Good leaders coach their staffs in ways that promote teamwork in prehospital care. They understand that it requires good judgment in addition to standard operating procedures to dictate responses to every issue. Judgment is an attribute best honed and perfected with the backup of a solid, caring, team-oriented system.

Attitude plays a vital role in safety. Proper attitude can be lifesaving. Yet many negative attitudes are so ingrained in the prehospital care industry that EMS personnel hardly notice them anymore. A switch to more positive, less dangerous attitudes requires organization-wide awareness and effort, The ripple effect is real, and as it reaches into the community, it increases community support for the EMS organization.

References/Endnotes

1. United Technologies ad, Wall Street Journal.
2. Jeff-J. Clawson and Kate Dernocoeur, K, Principles of Emergency Medical Dispatch (Englewood Cliffs, NJ: Brady 1988), p.35.
3. Pierce R. Brooks, "Officer Down, Code Three..." (Schaumberg, 111.: Motorola Teleprograms, Inc., 1975), p. 171.
4. Don Jackson, "Viewpoints: Paramedic Speaks His Mind," Dallas Morning News, March 14, 1984, and "Medics Report Backlash From Dispatch Case," Dallas Morning News, April 9, 1984.

CHAPTER 5
Personal Protective Equipment

Chapter Overview: Head to toe, there are garments and other personal protective clothing and equipment the prehospital provider should consider wearing to maximize personal protection. Even minor injuries may be avoided by using everything from proper protective headgear to shoes with ankle support and steel toes. Different types of body armor are discussed, to give the interested reader information on the safety advantages it can provide.

Lay people who work in the same setting from day to day might be able to take certain things for granted (such as a functioning elevator). Not so in EMS. Unpredictability and unreliability are part of EMS, and the problem itself is often based on the relative lack of safety of the setting: a fire, a collapsed embankment, fog on the highway or guns cocked and aimed. One way to balance the potential hazards lies with the important topic of personal protective equipment (PPE).

From head to toe, the EMS provider needs and deserves to have PPE. Decisions about what protective gear to provide is an important decision in each EMS organization. In some places, it is important to know how to stay warm and dry; in others, it is important to know how to stay cool. Readers must apply the principles offered here to local circumstances,

The "Magic Box" syndrome discussed in Chapter 2 also applies to the area of PPE. EMS personnel, in their haste to help others, have often been guilty of ignoring their own safety There are abundant photographs of EMS personnel on scenes, bareheaded, unjacketed and in flimsy shoes-when other emergency services workers at the same scene were using all their PPE. When the potential for harm exists, the smart EMS provider pays heed and uses available equipment to minimize that risk.

In the EMS firefighters realm, standards for protective clothing and equipment have been established. These include standards for helmets (NFPA 1972), gloves (NFPA 1973), footwear (NFPA 1974), protective hoods (NFPA 1971) and structural firefighting protective clothing (NFPA 1975). NFPA 1500, Standard on Fire Service Occupational Safety and Health Program, mandates that emergency personnel use this gear when in a hazardous area. Standards which detail performance criteria for protection against contamination of all types and for infection control practices and procedures have been compiled and published, including Standard on Protective Clothing for Emergency Medical Operations (NFPA 1999) and Standard on Fire Department Infection Control Programs (NFPA 1581). EMS personnel should not be left to their own devices to determine what PPE to use. EMS personnel subject to physical danger or performing in hazardous areas must be provided with (and must use) PPE that meets the published industry standards.

A head-to-toe survey of available protective equipment for EMS personnel follows:

Head Although EMS personnel tend to resist wearing helmets, anyone with experience with falling debris can vouch that wearing a helmet is worthwhile. Rescuers cannot control what falls from above or flies out during an extrication effort. But there is a way to control what it hits: a tough, replaceable synthetic shell is a better target than the skull and brain. Helmets are important protective gear in any situation in which head injury is a possibility They are also useful for identification of personnel during scene operations (particularly mass casualty situations) because they are easily seen and recognized.

Protective eyewear, whether prescription or not, is indicated when there is any splash or spray potential from blood or other body fluids. In addition, eyewear can protect the eyes from dust and other small particles of debris. These are frequently generated during extrication, by hostile weather and during helicopter operations.

Hearing protection is also important. One study of personnel in a busy urban system found that exposure just to sirens correlated significantly with hearing loss, accelerating it by 1 to 1½ times the rate expected for age-matched controls.[1] Because the long-term damage of high noise levels to hearing was not recognized for many years, many older emergency services workers sustained some hearing loss.[2] Now that the relationship is recognized, and with the knowledge that much noise exposure in EMS is uncontrollable, EMS personnel are urged to protect their hearing. This is a long-term quality-of-life issue.

Protection begins with a physical exam and audiometric testing to establish a baseline for hearing acuity, Also, one should consider using protective hearing devices (now mandated by some EMS agencies). In NFPA 1500, hearing protection is required for anyone exposed to noise in excess of 90 decibels (dB); for comparison, a normal conversation is conducted at about 65 dB. Earplugs can reduce noise reaching the inner ear by 15 to 26 dB; earmuffs achieve noise reduction levels of between 21 and 28 dB. Some models incorporate an earphone connected to the radio.[3] Hearing conservation programs are encouraged for local EMS personnel, and prototypes are now available.[4]

EMS personnel with long hair should tie it back, put it up or shorten it. Many organizations mandate this because long hair may be grabbed by an angry or panicked patient or may become tangled in equipment. For similar reasons, jewelry (especially dangling pieces) has no place on EMS calls.

Hands. Unprotected hands can be injured by sharp metal, glass and other common debris. Although wearing gloves can be awkward during hands-on medical care, they may provide some protection in other phases of care. They improve grip on the stretcher while moving a patient and can be helpful for other labor-intensive tasks. (The use of gloves to minimize exposure to blood and other body fluids is discussed at length in Chapter 14.)

Finger rings can cause a "degloving" injury, in which the soft tissue is pulled off the bone similar to the way a glove slides off. Such injuries have occurred when a finger ring snags on something like a nail when jumping from a high place.

Torso. According to NFPA 1500, "clothing that is made from 100 percent natural fibers or blends that are principally natural fibers should be selected over other fabrics that have poor thermal ability or ignite easily"5 Synthetic materials can be hazardous because they melt, drip, burn, shrink or transmit heat rapidly and cause burns to the wearer.[6] Uniform shirts and pants should be made of natural fibers or blends. These fibers also tend to be more comfortable in the temperature extremes commonly encountered by EMS personnel.

Wearing a badge on an EMS uniform is a common and long-standing practice in some places. There is potential for harm to EMS personnel who wear badges if they are mistaken for law enforcement personnel. People in crisis sometimes do not stop to read the badges before reacting negatively to them.

Bands of reflective tape are crucial on jackets, turnout coats and safety vests when EMS personnel are working in exposed locations, such as roadways. Citizen drivers are often so distracted by the excitement surrounding the incident that they fail to concentrate as they drive through the emergency area. Reflective tape provides improved visibility.

At the same time, situations occur where stealth and anonymity are preferable. For example, a coat with reflective tape at the scene of a barricaded subject might turn the wearer into a victim of target practice. At events like this, EMS personnel should be less visible while law enforcement officers secure the scene. One solution is to wear a reversible jacket. One side has plenty of reflective tape. The other is dark and promotes anonymity.

Outerwear should also protect the EMS provider appropriately from the effects of weather-particularly precipitation and cold. The materials chosen should adequately insulate workers and keep them dry. Some EMS personnel encounter all sorts of weather in the course of one shift: extreme heat or cold, rain, sleet, snow. Weather is constantly changing. Since it is never certain when a shift will end (for example, should a mass casualty incident or long transfer occur), it is helpful to be prepared. One EMS crew set out on a routine long-distance transfer and found themselves ill-prepared for the unexpected late spring snowstorm that suddenly arose. Then they ran out of gas and experienced some cold and unsure moments before the situation was resolved. Take extra warm clothes onto EMS units that operate in cold climates.

For day-to-day operations, versatile outdoor protective gear better allows EMS personnel to go in and out of harsh weather comfortably In snow country, an EMS provider may have to travel from a warm ambulance through the cold to a warm building and back through the cold several times a day. Yet that person must also be able to stay outdoors for lengthy periods of time, such as during an extrication. Levels of exertion fluctuate frequently and rapidly, from hours of waiting to intense physical work. The smartest way to dress for the elements in such an environment is in multiple layers: T-shirt, uniform shirt, vest or sweater, jacket. Wearing a hat and gloves will also aid in retaining body heat.

Lower body. Protective clothing for the pelvis, groin and legs closely matches that of the torso. There is one additional essential item of protective equipment usually carried on the belt: a portable radio. Since it is not always possible to predict the need to call for help, it is foolhardy to enter a prehospital scene without a portable radio. In one case, a paramedic crew was in a basement with a depressed and uncommunicative man. They kindly encouraged him to talk. Suddenly, the patient whirled into action, seized a heavy object and hurled it at the paramedics. They had no portable radio and couldn't call for help until they reached the ambulance-which fortunately is what they were able to do: The outcome could have been much different had they been trapped in that basement.

Another element of protection available to males is a protective groin cup. This is a personal decision. Although some do wear them, others feel the discomfort outweighs the risk of harm.

Feet. The importance of protective footwear should be easy to appreciate when considering the range of activity EMS personnel engage in: scrambling down embankments, climbing ladders and trees, lifting a heavy weight and carrying it safely past numerous obstacles. The safest all-around solution is to wear sturdy ankle-high shoes with steel toes. Good ankle support decreases the risk of turning an ankle; only one misstep with a loaded stretcher is enough to appreciate the need for it. Steel-toed shoes also minimize the chance of losing work time while nursing painful toe fractures, which can occur when heavy weights drop unexpectedly

Body armor. Whether or not EMS personnel should wear body armor is widely debated.[7] In some places, such as Los Angeles Fire Department, concealed body armor is provided and must be worn on responses to weapon assaults, shootings, sniper calls, domestic violence and police standbys.[8] In others, individuals are electing to spend their own funds for the peace of mind. As one paramedic said, "I wear my seat belt in a car, I wear my helmet on my motorcycle, and I wear my body armor on the streets." In his case, he said, the odds supported wearing body armor, considering that he has never had a crash in an emergency medical vehicle but has been shot at three times.

One argument in favor of using body armor is its ability to protect the wearer not only against bullets, but against other trauma as well. Although not specifically designed for this purpose, body armor has been credited with helping minimize or prevent injury from assaults with bottles, stakes, pitchforks and screwdrivers, as well as injury from jolts of electricity and lightning.' Body armor can minimize the impact of a steering wheel in a motor vehicle accident. In one case, it prevented impalement when the "A column" (one of the steel posts connecting the car's body and roof) collapsed into a medic's back while he was inside a wrecked vehicle."

An argument against using body armor is that it can instill a false sense of security. It is still possible to be severely or fatally injured while wearing body armor if an unprotected part of the anatomy is hit, such as the head. Serious blunt trauma can occur when the body armor absorbs the energy of a missile. Also, body armor can be uncomfortable to wear because it restricts movement and is hot; one study showed that only 25 percent of law-enforcement personnel who have body armor available wear it regularly."

Another argument against the need for body armor is that EMS personnel should not enter such dangerous situations. While this may be true, situations may not seem dangerous to the dispatching telecommunicator, or EMS personnel may get too far in before discovering a dangerous situation. Some scenes turn dangerous during the course of the call, even in the presence of law enforcement officers. Body armor is one measure that can provide the needed edge for survival. It is important to note that rural and suburban EMS personnel are just as vulnerable as urban EMS person-

Level	Deflects	Examples
I	Small-caliber, low-velocity handguns	.25, .32-caliber
II, IIA, III	Medium-velocity handguns and shotguns	.38-, .357-caliber 9mm pistol, 12-gauge shotgun
III, IV	High-Velocity, center-fire rifles and armor-piercing rounds	m-16 and AK-47 assault rifles

Adapted with permission from: Smith J: "Bite the bullet: A case for body armor." JEMS. 16(4), April 1991.

Figure 5-1 : Be sure to research body armor carefully and understand the differences in threat levels. Models provide different degrees of effectiveness.

nel and may benefit from wearing body armor. There are many examples of responders in rural areas who were spared critical injuries because they wore body armor.

Despite its deceptive name, body armor is not actually "bulletproof." It works by absorbing and distributing the impact of a ballistic missile. It is made of layers of an extremely tough synthetic fiber. Various degrees of protection (known as threat levels) are created according to the number of layers of the fiber (see Figure 5-1). The higher the threat level, the less comfortable the body armor, so at some point the individual has to decide where to trade off comfort for protection. There are two general styles of body armor:

• *Continuous-wear concealable:* The major advantage with this type of body armor is that it is on if it is needed. Proponents of concealable body armor like to point out that assailants who do not realize body armor is being worn will be less likely to aim at an unprotected area. Some models of continuous-wear concealable armor are wrap-around; those that do not are cooler. A reputable dealer should measure and custom-fit the armor for greatest comfort and mobility,

• *External wear:* Proponents argue that wearing body armor full-time is uncomfortable and almost always unnecessary because the risk of being shot is still considered low. External-wear body armor (commonly called a "tactical vest") is donned when a potentially hazardous situation is known to exist. When a situation is known to be risky, EMS personnel should not enter the area before it is secured by law enforcement personnel-and if the scene is truly secure, a tactical vest should not be needed. Unfortunately, it may sometimes be impossible to know when a scene is "safe enough." In situations of grand proportions, external-wear vests may make good sense; for example, they were issued to EMS personnel during the Los Angeles riots in 1992.

Like all equipment, body armor requires maintenance and care. Its ability to work properly may be diminished with wear and tear and also when wet (including perspiration). Periodic cleaning according to the manufacturer's instructions is important.

Summary

Making use of personal protective clothing and equipment is one way the EMS provider can enhance personal safety, even before leaving home. Using suitable material, proper shoes, gloves, a helmet--everything discussed in this chapter-will help minimize "nuisance" injuries-and may even be life-saving.

Like anything else, when it comes to safety, much begins with attitude. The mission of EMS is to care for people who are sick or injured; there is no reason the service provider should become sick or injured in the process. Using proper PPE can help.

References/Endnotes

1. Paul E. Pepe et al, "Accelerated Hearing Loss in Urban Emergency Medical Service firefighters," Annals of Emergency Medicine 14:5, May, 1985, pp.438-442.

2. Glenn H. Luedtke, ""Could You Please Speak Louder?": Coping with hearing loss in EMS,"JEMS, May 1988, pp.28-31.

3. Luedtke, p.31.

4. See "Fire and Emergency Service Hearing Conservation Program Manual," U.S. Fire Administration, 1992.

5. NFPA 1500, sections 5-2.7 and A-5-1.6, p. 390 and p. 398.

6. NFPA 1500, A-5-1.7, p. 398

7. This presentation of information is not intended as an endorsement of body armor, it is intended to provide accurate information to assist each local EMS agency come to its own conclusions.

8. Robert A. Ball, "EMS Under Siege: A Tale of Four Cities,"JEMS, April 1991, p.63.

9. Jim Smith, "Bite The Bullet: A Case for Body Armor,"JEMS, April 1991, p.30

10. Personal correspondence between the author and the survivor, 1992.

11. Smith, "Bite The Bullet, p.33.

SECTION II
Scene Operations

CHAPTER 6
Arrival and First Contact

Chapter Overview: This chapter contains information about considerations to be made during the early moments at an emergency scene. This includes assessing for environmental and interpersonal risks before exiting the vehicle, employing effective communications principles while making first contact with the people at the scene, and knowing how to cope with bystanders, relatives and potentially dangerous animals. Included, too, is a description of the Incident Management System, a process that can assist the prehospital provider in organizing any incident, particularly larger ones. The principles suggested in this chapter are intended to help minimize risk during the crucial initial minutes at an emergency scene.

Arrival is a period of great vulnerability The EMS team is visible to others but has not yet had the opportunity to completely assess the situation. Despite the advantages of obtaining advance information from properly trained Emergency Medical Dispatchers, EMS personnel face many unknowns when arriving at an emergency. Safety at a prehospital scene begins with thorough and candid evaluation during arrival. Without proper and attentive scene assessment, EMS personnel can get into trouble before they know it.

The first question to ask when arriving is, "Should we get out of the EMS unit?" Alert EMS personnel pose this question every time, even in circumstances that are apparently safe. Potentially dangerous situations may stem from traffic hazards, hazardous materials, crowds, questionable stability of structures, visible weapons, collapsing ditches and interpersonal violence.

In most cases, the arrival of help is desired and appreciated. However, there have been situations in which the patient's aggressors were unhappy to see medical assistance arriving. As one assailant once said, "If I'd wanted him alive, I would have shot him in the leg." Then he shot the patient again, even though police were at the scene. Be especially suspicious during the arrival phase even when responding to familiar addresses.

The EMS team should begin careful assessment while still in the EMS unit, where the retreat remains easiest; outside lies unfamiliar turf. On leaving the relative security of the EMS unit, EMS personnel face unpredictable and sometimes uncontrollable elements.

During arrival, do not back the emergency vehicle into the driveway. Doing so prevents thorough assessment of the scene, and provides extra time for someone with intentions of harming uniformed personnel to aim and shoot. If the call is on a dead-end street, the emergency vehicle should be driven past the scene and turned around; this way, it is facing the direction that allows for rapid departure if retreat is necessary. At night, leave cab lights turned off during arrival to prevent medical personnel from being "sitting duck" targets (see Figures 6-1A and 6-1B). Use a penlight or map light to record needed information (such as mileage or the time) or record it after the call.

In some EMS systems, the nature of the call dictates when to exit the EMS vehicle. For example, standard operating procedure in some places is to await law enforcement backup on all reported shootings and stabbings, In other places, EMS personnel are left to decide on a case-by-case basis whether they feel comfortable exiting the EMS unit.

Sometimes, police are not available to protect EMS responders. Perhaps the need for them was not evident to the telecommunicator, so they were not included in the initial emergency response. Even where relations between EMS and law enforcement are excellent and police backup arrives quickly, fiscal constraints may sometimes leave police struggling to handle their own call volume. On calls that require both police and EMS (such as motor vehicle crashes and domestic violence), law enforcement may not respond as rapidly as EMS. In rural communities, lengthy response times by law enforcement personnel may often be due to distance. In some places,

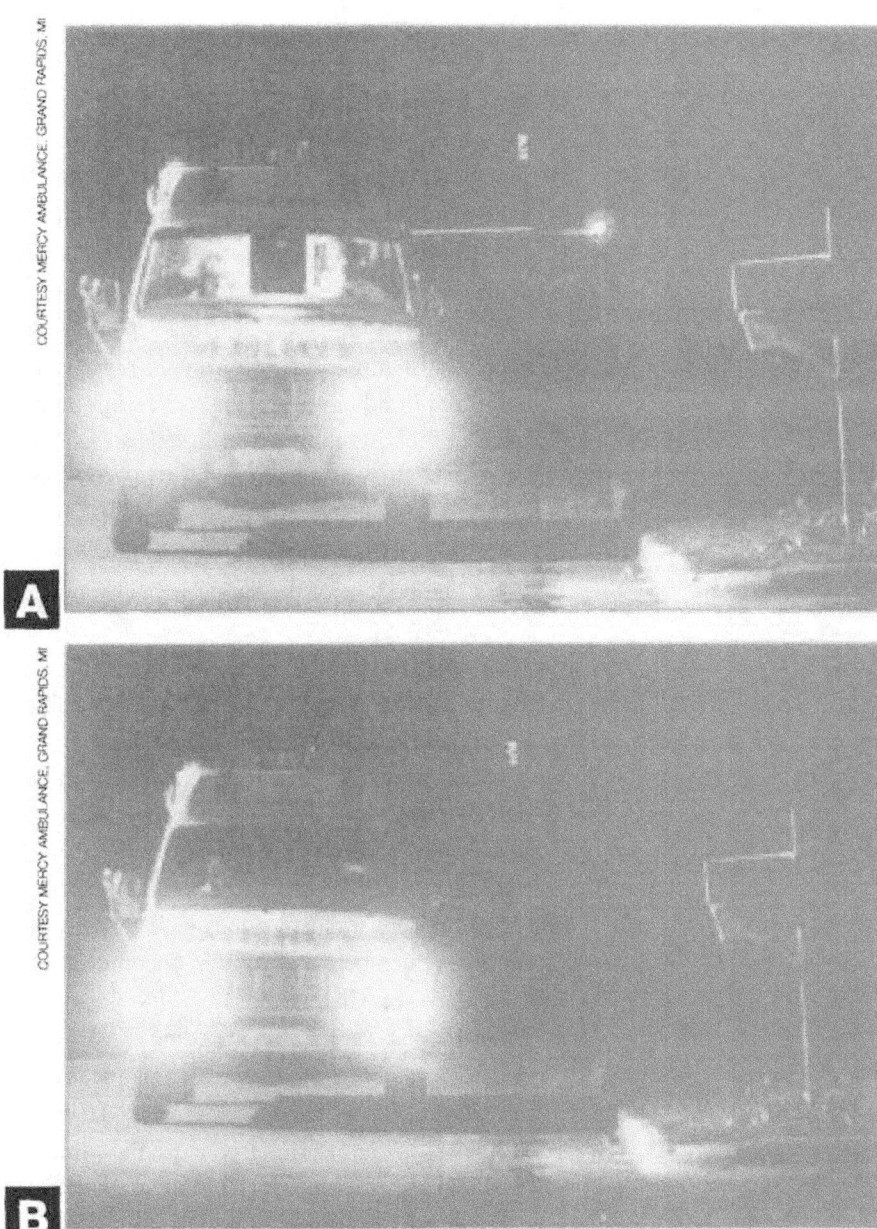

Figure 6-l: During arrival at night, keep cab lights off to minimize the chance of being an easy target. Note the difference from outside when the cab lights are on (photo A) and when they are off (photo B).

EMS workers know, for various reasons, that police support might not arrive at all.

When a scene is dangerous, the EMS provider is justified in leaving. Suppressing dangerous situations is a law enforcement task, not that of prehospital emergency medical service personnel. The most important principle is to get away from the area. Retreat and wait for law enforcement officers to secure the scene-no matter how long it takes. These actions may need to be justified later- but at least the individuals will be alive to do so. (EMS personnel are usually found to be completely justified in retreating.)

While approaching on-scene civilians, "read" their body language on three levels: their overall body language, their facial expressions and their eyes. Stance and posture create an initial impression. The most outward emotion is often overtly demonstrative. The facial expression may sometimes contradict the overall body language. Finally, the eyes, as a direct sensory link to the brain, can provide extremely valuable information.

Conflicting body language may mean various things, Overall body language that demonstrates extreme excitement, a facial expression that is angry and eyes that show fear may all be accurate reflections. Try to discover why It might turn out, for example, that the person is the father of an injured toddler. The man's overall body language is hyperexcited because of a buildup of adrenaline. His face looks angry, and he uses angry words: "What took you so long?" The true source of anger is at himself for leaving the toddler's stair gate open while he started dinner. And the fear (the truest emotion) revolves around whether or not his son, lying in a heap at the bottom of the stairs, is badly injured.

First Contact

Optimally, the EMS team would always be able to evaluate scenes for safety from the security of the EMS vehicle. But this is not usually the case. The EMS provider must leave the EMS vehicle and approach the situation on foot. It is often obviously safe to get out and make first contact with those at the scene. Maybe the patient is on the street or sidewalk, surrounded by other emergency providers. Perhaps the police have already arrived and are indicating it is safe to proceed to the patient. (When getting out, remember to check for oncoming traffic in order to exit the EMS unit safely-see Chapter 2.)

First contact with those needing EMS is hugely variable. From loud, boisterous streets full of excited people to the too-quiet, nobody-at-the-door scene, this aspect of emergency response is always uniquely challenging. It is another time of extreme vulnerability Use a high degree of suspicion and open every sense (including the gut feeling or sixth sense) to cues about the scene's safety.

If other emergency personnel are present, the safety margin is enhanced both by their numbers and by the fact that they may already be providing care. Although others have already presumably assessed the scene, each newly arriving individual is wise to evaluate the scene for himself. The others may have entered the situation without proper attention to detail. Subtle hazards may have gone unnoticed. Continue 360-degree assessment for physical hazards (hazardous materials spills, fire, downed power lines, etc.), structural hazards (unstable footing, weak or shabby stairs, etc.) and interpersonal hazards (unstable individuals).

When an EMS unit is first to arrive at a scene, extra caution during size-up is appropriate, since the first personnel to arrive tend to face more unknowns. Subsequent rescuers can usually benefit from the safety assessment efforts of those who arrived earlier. Upon approaching the scene, each rescuer should act as an "information vacuum"-taking in every par-

Sidebar 1:
Acting the Part: Choosing Roles for Each Scene

The EMS provider must practice assessing situations so that choosing an interpersonal approach becomes second nature. Some situations demand a hard-line, authoritative approach; others work better when the approach is softer. Between these extremes is a continuum of interpersonal approaches (see Figure 6-2). EMS professionals who choose to learn how to move up and down along this continuum will find themselves working effectively in a broader range of circumstances. For example, a large, authoritative-looking body-which is great in a bar frequented by unruly people--can reduce a child to tears of fear and generate defensive, upset parents. Such a person can learn to appear softer. Conversely, a petite, gentle presence may be perfect for dealing with a child but will earn no respect at a tough bar.

Begin by determining inherent interpersonal style, which is that point where each person naturally lands along the hard-to-soft continuum. Inherent style depends on stature, size and other physical attributes, as well as habitual vocal dynamics, facial expressions, "presence" and word choices. Some people naturally seem tough and challenging; others seem timid and mild. The majority of experienced EMS personnel tend to be somewhat tougher than the average lay person, since they tend to be "take charge" people.

Learning how to soften and toughen one's approach is one way to learn how to defuse-not escalate-intense prehospital situations. To rely on one single approach is impractical. Although there may seem to be little people can do to change what they look like, a naturally mean-looking face can be trained to look more compassionate, and someone with a meek-looking face can create a more authoritative expression. Overriding one's natural expression is both possible and worthwhile. One paramedic did this by investing time retraining the muscles in his face to look friendlier-and it worked.

Soft	Medium	Hard
Characterized by gentle voice, kind facial expressions, non-angular body language.		Characterized by powerful and controlling body language, including stern facial expression, rigid, angular stances, and louder, more firm or harsh tones.

Figure 6-2: This continuum is a building block for improving one's ability to react appropriately to each field situation. Some EMS personnel are control-oriented people, and are naturally somewhat to the right of the medium. However, nowing various approaches to try provides versatility when dealing with strangers who are in crisis.

ticle of information that will assist in decision-making about the scene.

An early concern during approach to the patient is whether the patient is holding a weapon. When summoned for a "person down," the EMS team's main task is determining whether there are underlying injuries or illnesses. Sometimes, that person is simply resting in public. Homeless people know they are vulnerable and often sleep while holding something that could provide self-defense if necessary It may not be a traditional weapon such as a gun or knife. A sharp bottle opener or other improvised weapon is potentially harmful if the EMS provider does not properly anticipate it. Use extreme caution when first contacting someone until both hands have been checked.

Do not startle resting or sleeping people. EMS personnel sometimes cannot tell immediately whether a person lying down in public is just sleeping or is truly injured or sick, Understand that some people may react violently to being awakened abruptly by loud noises, noxious odors or being touched. The EMS provider should use a stance which allows rapid retreat in case the patient wakes up quickly and violently. Never simply lean over at the waist. Keep one foot forward and most of the weight on the back leg. This leaves the ability to push back quickly if necessary (see Figure: 6-3).

Effective Interpersonal Communications Tactics

One important decision for the EMS provider to make early in the encounter is which interpersonal approach will allow rapid development of trust. This decision can be easier if the right information is gathered during the approach and first contact. Trust is earned when an EMS provider discovers the best way to achieve good rapport with the strangers who called for help. Rapport-building is an important safety measure, since people are less inclined to lash out at those they trust. Remember, even people already known to the EMS provider can be changed by the circumstances of an emergency medical event.

The outcome of many interactions depends on the willingness of both sides to extend an effort. It is possible to gain effective control of most chaotic situations through the use of good interpersonal communication. It is also possible to escalate potentially explosive situations by using inappropriate communication. EMS personnel who use the correct skills can powerfully influence how well others cooperate with them.

The EMS provider can develop a variety of communication strategies that will work in assorted settings (see Sidebar 1: "Acting the Part: Choosing Roles for Each Scene"). Meeting the diverse needs of the public by honing the ability to assume different roles is part of the art of a safe EMS experience. The combination of all the elements is summed up by the concept of presence. Good scene presence is a professional demeanor which encourages others to respond positively and willingly It indicates the ability to work well with people. As one self-defense quote goes, "The fastest draw is when the sword never leaves the scabbard, the strongest way to block is never to provoke a blow, and the cleanest cut is the one withheld."

Many elements of communication can add to an EMS providers interpersonal versatility These include:

• *Understanding the concept of interpersonal space* (see Figure 6-4). EMS personnel often rush headlong into what is known as "intimate space" in order to check a pulse or hold a head and spine steady Some people may regard this as an intrusion and become stiff and defensive. Minimizing negative reactions may be as simple as moving out to the next zone. Build trust by taking the time to achieve some rapport. The patient care provider should explain why EMS has arrived, introduce himself (by name and function) and express an interest in resolving the situation to everyone's satisfaction. Once the person relaxes, entry to inner zones is both easier and safer,

• *Positioning relative to another person's eye level.* Towering over someone

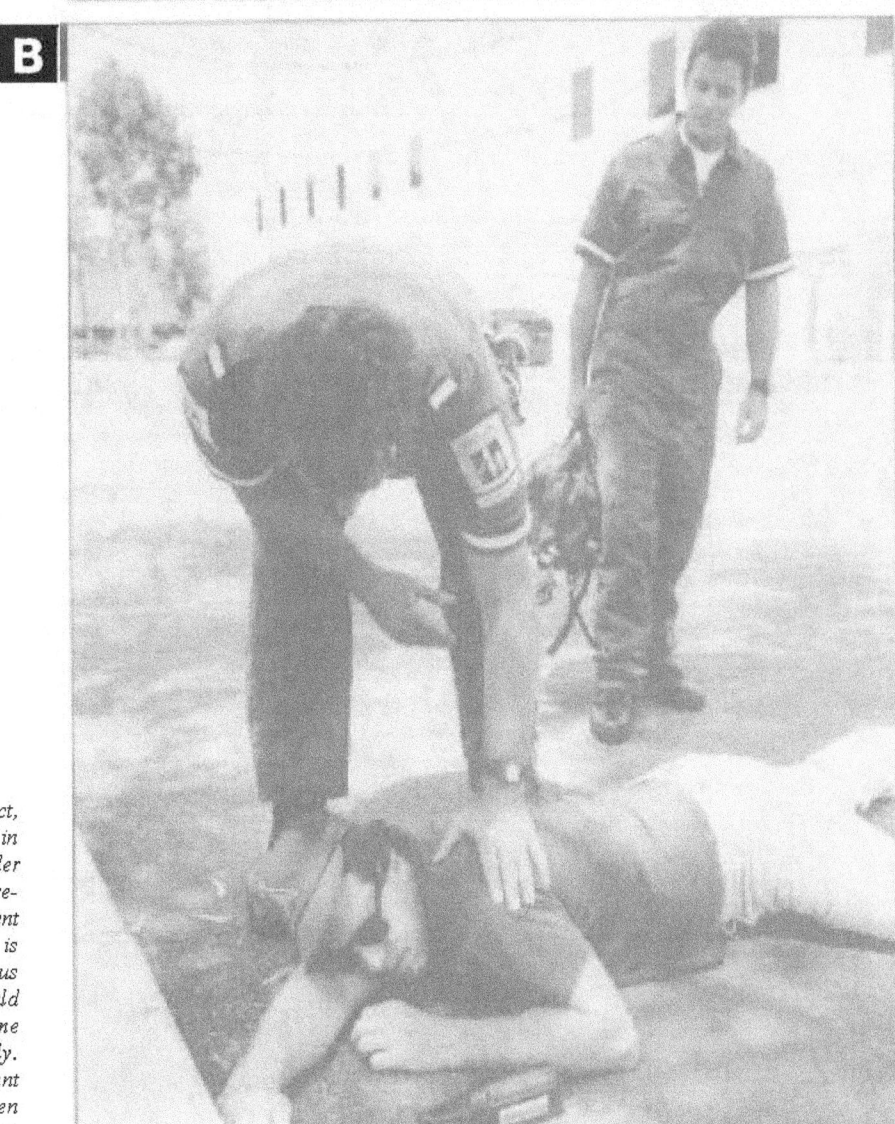

Figure 6-3: During first contact, be extremely cautious. Note in photo A how the EMS provider has his weight back, and is prepared to push away if the patient comes up quickly. His partner is also watching for suspicious behavior. In photo B, he would have a much more difficult time backing off quickly and safely. Remain especially observant until both hands have been checked for weapons.

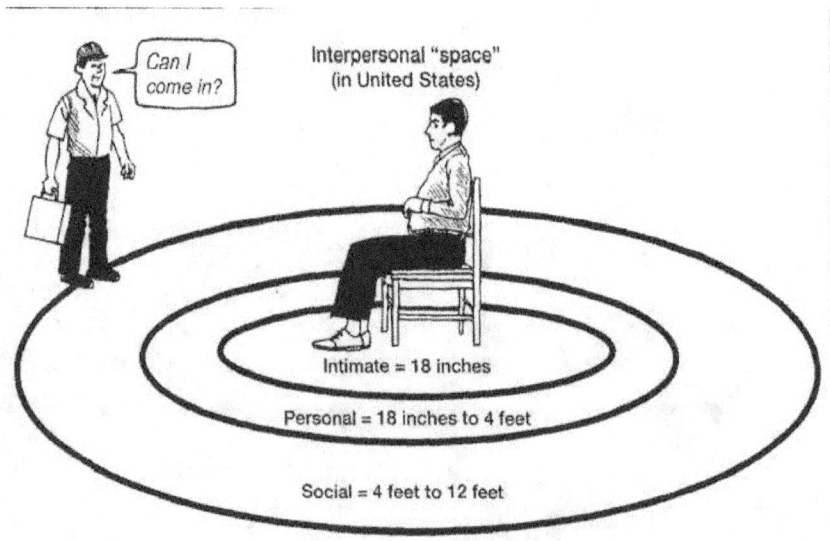

Interpersonal "space"
(in United States)

Can I come in?

Intimate = 18 inches

Personal = 18 inches to 4 feet

Social = 4 feet to 12 feet

Public = 12 feet or more

Adapted with permission from Demacoeur K: "Streetsense: Communication, Safety, and Control" 1990, p. 64; Prentice Hall Englewood Cliffs, NJ.

Figure 6-4: These interpersonal zones are typical in North America. The distances appropriate to each zone may change with people from different cultures.

creates a commanding effect. A neutral, straight-across position demonstrates a willingness to work together. To be below the patient's eye level is to impart a sense of control to the patient. Each has its appropriate use.

• *Effective use of physical stance.* A square stance appears tough, strong and authoritative. When a tough stance is most appropriate, be careful to give simply the illusion (by positioning the hips and shoulders squarely) while positioning the feet one in front of the other (see Figure 6-5). This provides the opportunity to push back should retreat become necessary. A softer stance is created by putting one foot ahead of the other and opening the angles of the hips and shoulders (see Figure 6-6). Arm position completes the picture. For a tougher-looking stance, the arms may be folded or on the hips. To appear more willing to negotiate and settle things peacefully, hands that are open with the palms down send a conciliatory message. Never put hands in pockets or fold them tightly in case they are needed suddenly for balance or to block a striking blow.

• *Understanding the power of word choices and tone of voice.* Volatile situations can often be calmed through appropriate word choices. For example, an EMS provider might say, "I can see your arm is hurt. Let me come over to you and help stop the bleeding, okay?" rather than "Look at your arm! You're bleeding! Can't you see that? Don't you know we're here to help?" The latter option is belittling and does not indicate what the EMS personnel intend to do about the problem. Avoid authoritative or demean-

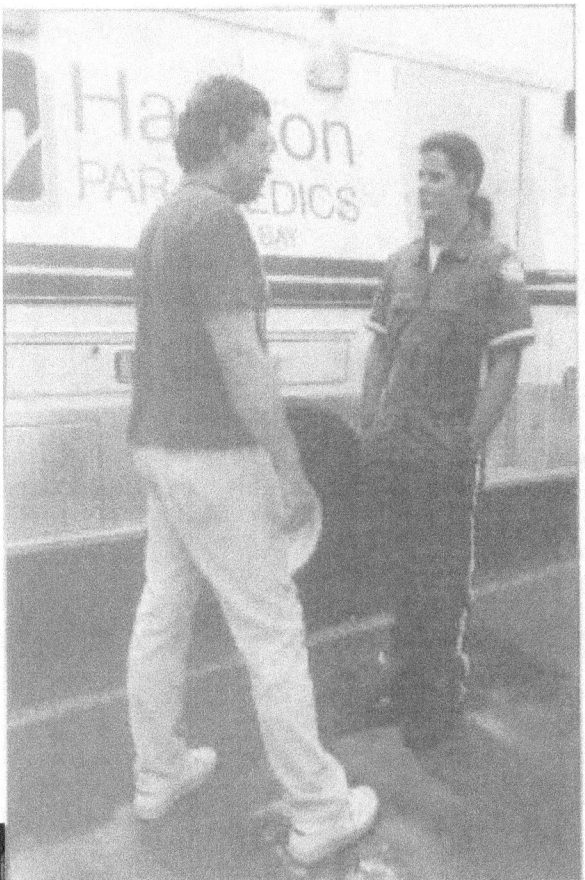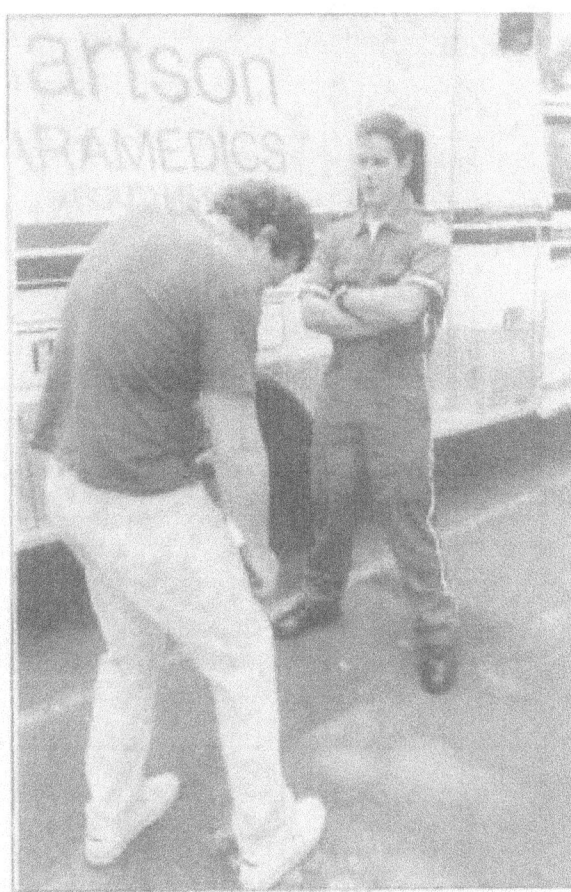

Figure 6.5: Standing with the feet side-by-side and with hands in pockets (photo A) or with arms tightly folded (photo B) is dangerous. It would probable take too much time to position hands to deflect an attack, and it is nearly impossible to back off quickly without one foot forward. These postures may also signal unintended lack of compassion or too much authority, which may irritate an escalating situation.

ing word choices and tones of voice.

• *Verbalizing.* Tell others what is being done and why. This will avoid having a medical procedure perceived negatively by the lay persons at the scene. By saying, "This splint may hurt while we're putting it on, but once it's there, the leg should feel better, and will help avoid further damage," cooperation, trust and rapport are all strengthened.

• *Requesting feedback.* Involving others helps them feel part of the solution-which can eliminate or minimize the "us versus them" mentality which can be dangerous to field responders.

Congruence of all the elements of communication is important. For example, a commanding voice and tough stance are not convincing if the face looks timid or frightened.

Certain irritating mannerisms can quickly push some people into aggressive responses. Examples include leaning against a doorjamb, tapping one's palm with a pen, pointing a finger at someone's face, rolling the eyes or laughing inappropriately at the things people say. EMS responders should avoid looking disinterested, bored, or brutal. A disrespectful or discourteous demeanor can also be dangerous. Different mannerisms and responses will irritate different people and do nothing to enhance a posi-

Figure 6-6: An open stance can pacify a situation. Note the open hands (which are nonetheless positioned to deflect a blow!) and the softening angles of having one hip and shoulder back and the head tilted.

tive public perception of EMS.

Incident Management System

In EMS, a scene without adequate leadership can deteriorate quickly. One approach being adapted to the EMS environment is the Incident Management System (IMS). IMS is an outgrowth of the Incident Command System (ICS). The National Fire Protection Association (NFPA) *Standard on Fire Department Incident Management System* (NFPA 1561) states that "the purpose of an incident management system is to provide structure and coordination to the management of emergency incident operations in order to provide for the safety and health of. . . persons involved in those activities."[1] It is a process for creating order from disorder through a comprehensive approach to leadership and task delegation. Versions of IMS have been used for handling wildland fires, urban fire scenes, hazmat situations and everyday EMS scenes.

Anything that promotes effective, consistent scene leadership is a key issue in prehospital safety No matter how an EMS system is configured, or how many responder tiers exist, this could be routinely achieved through use of the Incident Management System. The primary principle of IMS is for one leader to make scene management decisions, "In cases where an identified chain of command is not identified . . . problems emerge. This was emphasized by George Roderics, former director of the Mayor's Command Center, Washington, D.C., after the Air Florida crash of January 1982. '[If] there is no leadership, there is no direction, there is no coordination. And so, it does it all by hit or miss. And if it hits right, we're lucky If it misses, we hurt . . . Many disasters falter in the early moments or the first hour while people try to sort out who's in charge.' "[2]

Prehospital scenes, no matter how large or small, are more orderly and organized when there is one incident leader. Decisions are based on the big picture rather than piecemeal as new information becomes known. This improves incident coordination even on two- or three-person crews handling routine emergency calls. The person assuming the leadership role should be prepared to take on certain functions, delegating them as needed depending on the circumstances of the call. For example, an available first responder might be asked to scout out the best path from the scene to the EMS vehicle.

The incident commander should be the individual with the best ability to assemble the resources necessary and macro-manage the scene from an overview perspective. Roles should be clearly delineated in standard operating procedures, The information must be developed and shared across jurisdictional lines for times when multiagency responses occur.

Developing a well-rehearsed, well-thought-out command system can have positive consequences. Although especially useful for large-scale or mass-casualty situations, a level of IMS can and should be implemented on every EMS scene. Doing so has several benefits:

• *Better scene control.* One person determines priorities and attends to or delegates them. Consolidated action plans among local responding agencies with cooperative application of strategies and tactics allows the common goal-good patient care-to be achieved.

• *Effective leadership.* Without it, scene safety can be lost (or never gained). Predictable command structure helps increase confidence among EMS personnel because one in-charge person is responsible for overall safety. When no system is in place, efforts to address safety are more hit-or-miss.

• *Practice.* EMS personnel get accustomed to the principles of IMS when they become everyday routine. The system is then an instilled habit when low-level multi-casualty incidents occur, such as multiple shootings

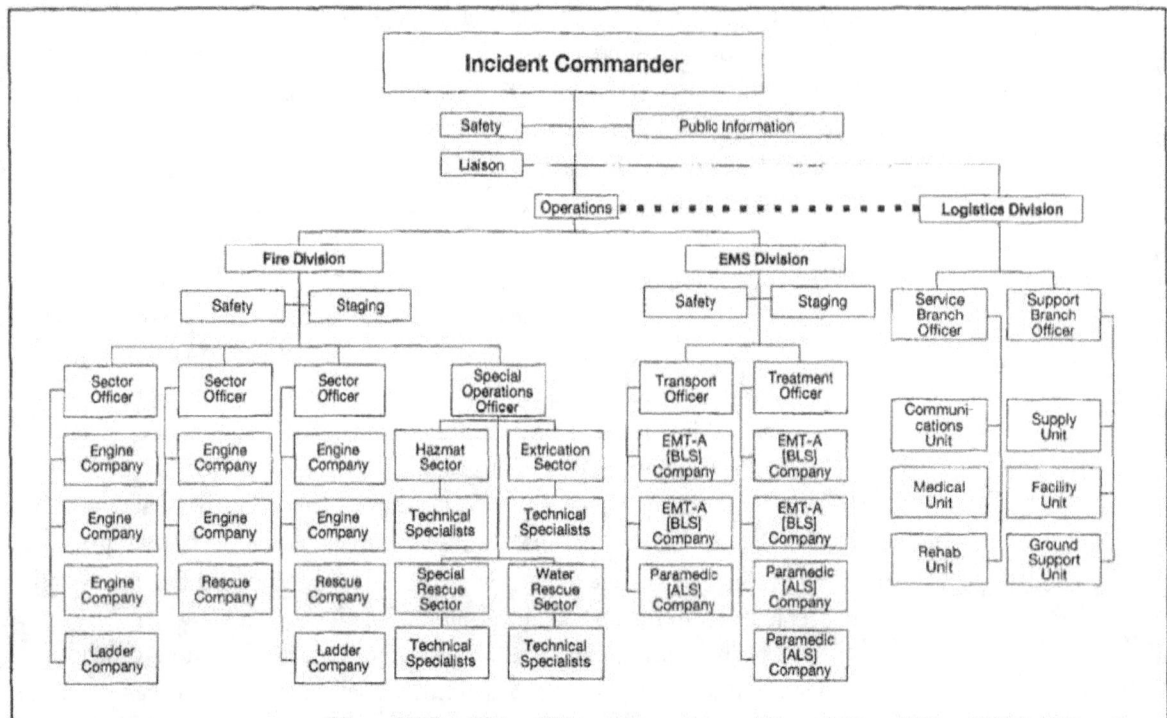

Figure 6-7: Multi-branch emergency incident management system organization flow chart

and motor vehicle crashes involving several vehicles. In the less-likely event of a large-scale mass-casualty situation, all local EMS personnel are better prepared to follow a prescribed model for organizing and handling the situation.

• *Consistent size-up and coordinated action plan.* An initial determination is made as to the amount of equipment and manpower that will be needed to control the incident. Size-up continues as conditions change and more information becomes available. Equipment is ordered and sent to a common staging area. Even if just one additional emergency vehicle is needed, it should be told which route to use to access the incident and where the staging area will be.

Even EMS personnel with years of experience who may have developed personalized scene control habits will find the basic framework of IMS appealing. Some of the specialized components (often referred to as sectors, groups or divisions) of an EMS Incident Management System that may be used at larger incidents are (see Figure 6-7):

• *Staging.* In a strict sense, this is grouping of resources in one area. One member is designated Staging Officer to coordinate the availability and use of these resources. For the EMS provider, this is creating an optimal setting for delivering appropriate, safe patient care. Equipment is brought to a forward staging area to assure quick and efficient action, while transport vehicles and other apparatus remain at the main staging area some distance away

• *Triage.* This is the process of sorting patients so that the greatest good can be provided to the greatest number. Typically, a Triage Officer is established early in an incident to coordinate the initial assessment of patients and report to the Incident Commander on the number and severity of injured or ill persons. The IC can then determine the necessary resources based on this information.

• *Treatment.* Patient care. The Incident Commander for everyday EMS calls may also be the person attending to the patient. When the number of patients or complexity of the incident increases, however, the Incident Commander will have too many responsibilities to provide patient care. The IC may not even be able to directly supervise patient care. At these incidents, a Treatment Officer is appointed. Note that safety and patient care are concurrent concerns; however, when a threat to safety occurs, it is the higher priority, Treatment might need to be suspended temporarily. For example, it may be necessary to load the patient and drive away, stopping (if necessary) when a safe environment is assured to initiate essential treatment. With this in mind, it is important for the Incident Commander and Treatment Officer to work closely

• *Transport.* Transport can include both moving the patient to and from the EMS unit and transport in the emergency vehicle. It also includes deciding where a patient should be taken. For this reason, the IC assigns a Transportation Officer to coordinate this information. (Everyday safety considerations related to transport were presented in detail in Chapter 2.)

• *Safety.* At multi-casualty situations, one person should assume the role of safety Officer. At everyday scenes, all EMS personnel must attend to safety considerations. For example, if one person notices a hole in the embankment on the way down to the crashed car, pointing it out may avoid injury to others.[3] (refer to Chapters 15 and 16 for more information on the Manager's role in Safety and Safety Officers.)

Bystanders and Relatives

Bystanders and relatives pose a particular challenge. They, too, are in a state of crisis or upheaval. Although many bystanders and relatives maintain their composure, others are dangerous. People watching someone else having a medical emergency may react in an unhelpful manner out of ineptness, hysteria or even malice. Their actions may stem from the media's portrayal of EMS; Hollywood's version sometimes sends the message that a person should be overwrought and defensive. Some bystanders are overly excited because they do not understand the capabilities of their EMS service. They may shout, "Just get him to the hospital! Go! Go to the hospital right now!"

As with patients, it is important to establish rapport and build trust with those bystanders and relatives who will be involved with the emergency, Most are a legitimate part of the scene, and many have valuable information that the patient may be unable to provide. During initial contact, the EMS provider can defuse many powerful emotions by demonstrating a presence that broadcasts the message: "We are in control. We can handle this situation." The scene leader should exchange introductions and verbalize what is happening as the scene develops or delegate the task to someone on the prehospital team. Tactics that calm a scene also increase its general safety.

Dangerous Animals

A team of paramedics was once dispatched for a patient with a complaint of overdose. On the police channel, they heard the police dispatcher calling for backup law officers. Multiple cruisers responded with the intensity usually associated with a shooting. The paramedics arrived to find the police awaiting them.

"There's a *really* nasty Doberman that lives at this address," they explained. "We didn't want you going in without us." (This underlines the value of good interagency teamwork through positive relations and mutual commitment!)

Domestic animals sometimes pose a safety hazard to EMS personnel. The presence of any animal should be regarded seriously. Even placid animals may become anxious during a medical emergency. When possible, enclose animals (particularly dogs) elsewhere on the premises, regardless of how well the animal seems to be coping with the situation. Delegate this task to someone who knows that animal, if possible. If no such person is available, the animal control officer should be summoned to deal appropriately with the animal.

Retreat from a violent or dangerous animal may be possible if the situation becomes obvious early enough. Be aware of the location of potential protective barriers (such as the EMS unit, a fence or a door) while entering the scene.

In some situations, it may be necessary to control upset or vicious animals. Although humane treatment should always be a consideration, there may be times when a life-threatening situation justifies harsh measures. Options available to EMS personnel may include using a blanket as a crude snare for a small animal, flashing a high-candlepower flashlight beam in the animal's eyes, or momentarily stunning the animal with a short blast from a CO_2 extinguisher or fire hose. Each of these is solely intended to allow time to reach a protective barrier. They are *not* intended to overpower the animal for the purpose of continuing with emergency medical care in the animal's presence.

Additional tips when facing a hostile animal include:

• *Try not to let fear show*. Animals can sense fear. Some respond aggressively to it. EMS providers who are not adept with animals can diminish their fears by spending time around animals and asking animal lovers to educate them about safe and effective ways to handle animals.

• *Adopt a "take charge" presence*. Who is in charge? The animal? Or the humans who are trying to help someone who is sick or injured?

• *Use a commanding, loud tone of voice*. Sounding authoritative may fool the animal into complying with orders to stay back. Use words an animal might recognize. Try "Down!" or "No!"

• *Do not turn away from the animal*. Watch the animal. Back off slowly, and hold anything available (such as the medical kit) between yourself and the animal. At the same time, never try to stare down an animal. Some perceive a steady stare as a nonverbal challenge, and will respond aggressively. This is especially true of certain dogs, and also hooved animals, such as bulls.

• *If attack becomes inevitable, try to protect the throat and face.* Expose or "offer" a less vital part of the anatomy. Injuries may be significant, but at least they may not be life-threatening.

Dogs are not the only dangerous animals EMS personnel have encountered. One EMS crew was sent to a motor vehicle crash and found an overturned van that had been transporting venomous snakes, including a cobra. Farm animals-including bulls, hogs and goats-can also inflict serious injuries. In addition, some people keep exotic animals as pets. When confronted by such situations, await the assistance of a trained and properly equipped animal handler. For example, EMS personnel should never feel obligated to enter the cage of a dangerous animal at the zoo until the animal has been securely restrained or relocated.

Summary

The vulnerability of EMS personnel during the time of arrival should never be underestimated. This period has the greatest number of unanswered questions and the most unknowns. The EMS provider needs a great deal of information very quickly, generating an intensity that often threatens to override more deliberate inclinations.

Arrival must be a time for concentration and attention to the unfolding scenario, through use of every information-gathering mechanism available. The skeptic that lives within each person must be fully alert, as should each of the five senses.

The Incident Management System provides for an organized approach to scenes. It creates a structure or framework for the EMS provider to implement leadership, establish accountability, and successfully complete the service of EMS.

References/Endnotes

1. NFPA 1561: Standard on Fire Department incident Management System, 1990 edition, (Quincy, MA: NFPA, 1990). Section 1-2.2, p.1561-5.
2. Chris Eldridge, "ICS: The Greatest Good for the Greatest Number," The Gold Cross: The Magazine of the New Jersey First Aid Council, Winter, 1991, p.16.
3. Kate Dernocoeur, Streetsense: Communication, Safety and Control, 2nd edition (Englewood Cliffs, NJ: Brady, 1990). See Chapter 5: Death and Dying.

CHAPTER 7
Outdoor Operations

Chapter Overview: This chapter discusses safety issues specific to outdoor settings. It includes recommendations for working at roadway scenes, safe places to park, use of warning devices such as cones or flares, and other methods for controlling traffic. Safety-related principles of disentanglement and approaches to stopped vehicles are examined, and outdoor environmental hazards, large preplanned events and how to gain cover or concealment from threats are addressed.

Outdoor scene operations pose distinct safety hazards. EMS personnel have less control over the surroundings. The weather is a more prominent factor. There may be a larger area for a crowd to collect. Roadways are unpredictable and dangerous. In a few cases, outdoor operations make EMS work easier, such as when the patient is located next to a place to park the EMS unit. In these cases, transition from the turf controlled by strangers to the relative safety of the EMS unit can be very rapid. Otherwise, being outdoors can be a far different experience from being indoors, where other safety considerations may occur (see Chapter 8).

Roadway Operations

EMS scenes that occur on or beside roadways can be extremely hazardous; many emergency responders' lives have been lost in the roadside emergency care environment. Ideally, EMS personnel could always rely on the police to properly provide traffic control, but this is unrealistic. In many areas, assistance is either slow to arrive or unavailable. EMS personnel need to know methods to maximize their own safety and the safety of those they serve.

Most EMS responders want to trust the general public not to cause them harm. But imagine the following scenario:

It is nighttime and raining lightly-enough to smear an old, pitted windshield when the well-used wiper blades have just been turned on. The driver's blood alcohol level is .04-not legally drunk, but not sober either. Looking like the aurora borealis, the emergency lights on the three emergency vehicles at the scene up ahead mesmerize the driver. Busy focusing on the hubbub ahead, the driver fails to notice the EMS provider who is standing 100 yards from the accident trying to slow and direct traffic.

There is more of a chance of being hit when EMS personnel are not constantly aware that other drivers may not have their best interests in mind.

Basic areas of concern addressed in this section are where to park the EMS unit, working near moving traffic, proper use of warning devices, basic safety principles of disentanglement and how to approach a stopped vehicle. Local protocol that varies from the suggestions made here should prevail if it provides for a safe working environment for EMS personnel.

Parking the EMS unit

During arrival, while the EMS crew can still survey the scene through the windshield, the driver must choose the best place to park. Considerations include:

- which spot is safest
- which is most convenient for patient care
- which will best facilitate traffic flow around the scene
- which allows the best outlet for departure
- whether other emergency personnel are already on the scene or will arrive later

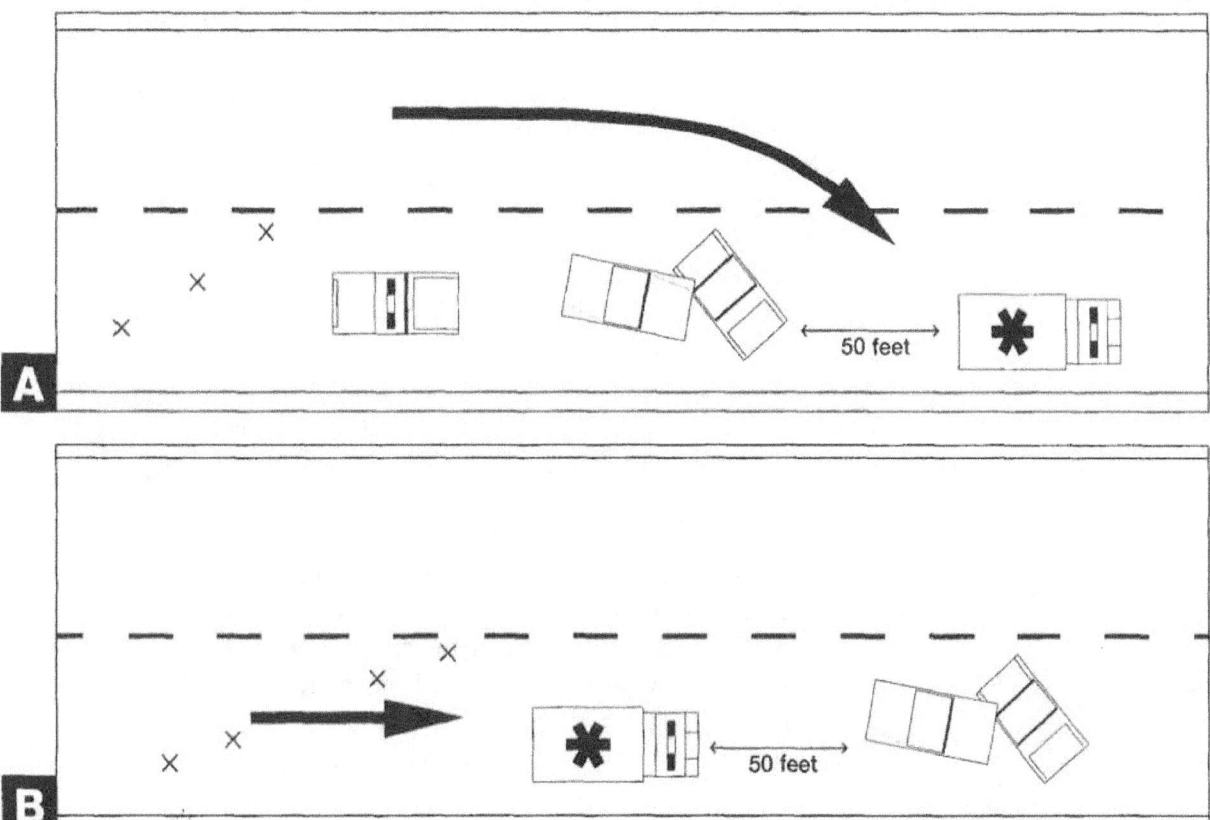

Figures 7-1A and 7-1B: When positioning options are limited, remember to ensure that a vehicle is parked to create a buffer zone for on-site personnel. Whenever possible, this should be a vehicle other than the EMS unit, so that the unit can be moved ahead of the crash site for better and safer access to equipment and the patient compartment.

Of these, safety must prevail. There are several considerations. Some people believe it is safest if the EMS unit is placed between oncoming traffic and the involved vehicles. This provides emergency warning light at that end of the crash site, illuminates the scene with the EMS unit's headlights, and provides a physical barrier from approaching traffic while operating at the scene. However, once the patient(s) and EMS personnel are in the EMS unit, they are endangered by being closest to oncoming traffic during preparation for exiting the scene. Therefore, while it makes sense to position an emergency vehicle at the end of the crash site, using the EMS unit for this is probably not the best choice if another emergency vehicle is available. It may be better for the EMS unit to be moved ahead of the wreckage when another vehicle arrives to create a barrier against traffic (see Figures 7-1A and 7-1B). Setting the emergency brake wherever the unit is parked may help reduce vehicle movement should it be struck.

During arrival, EMS personnel must also note such elements as spilled fuel or other hazardous materials, downed electrical wires, broken glass and injured people on the pavement who might have been hit by or thrown from the crashed vehicles. Park far enough away from the danger zone to ensure the safety of the EMS unit should a fire or explosion occur. One reference recommends a zone of 50 feet in all directions from a "normal" wreck and 100 feet or more where fire or hazardous materials are present.[1] Certain hazardous materials situations will require much greater safety zones-sometimes as much as half a mile.

Always park uphill and upwind from sites with potential for ground or air contamination. This prevents the material from washing into the tires of the EMS unit and hazard-laden air from endangering personnel or contaminating the inside of the EMS unit.[2]

Look at the terrain when deciding where to park. If the incident is on a curve or at the crest of a hill, there is little to prevent other drivers from approaching the site at full speed until warning devices are set out. Where to park the EMS unit in these instances depends on whether backup emergency assistance is close to arrival. It may be necessary to place the EMS unit by the curve or at the crest of the hill until another emergency vehicle can take its place. Then it should be moved to a more protected and convenient location.

Completely avoid parking across from the incident site with a lane of traffic between; someone could be hit while crossing back and forth to the EMS unit. If the EMS unit has approached the site from the opposite direction, it may be appropriate to park on the "wrong" side of the road, facing the scene. This allows EMS personnel to light up the scene with the headlights while avoiding other moving traffic when on foot. Be careful not to blind oncoming drivers with emergency vehicle headlights or scene lights.

Moving traffic

Although the emergency is foremost in the minds of those involved, it is often viewed as a major annoyance to other motorists. When the pressures of modern society are compounded by the traffic tie-ups associated with crashes, motorists have been known to take extremely dangerous risks, such as driving the wrong way on the highway to bypass obstructions. Many are unconcerned with the safety of emergency personnel. Thus, safety risks due to exposure to moving traffic should be the personal concern of every emergency responder.

Traffic can be blocked in various ways. The entire road may be impassable. There may be one lane available for bidirectional traffic. On multi-lane roads, several lanes may be useable. Sometimes, traffic is not obstructed, yet slowdowns still result from the curiosity of passing drivers. When law enforcement personnel are unavailable for traffic control, EMS personnel should employ the following principles for safe and effective direction of traffic:

• *Create a buffer zone with one of the emergency vehicles.* Parking the vehicle between moving traffic and the crash site is a good idea. Offsetting the EMS unit slightly will leave a small "corridor" for EMS personnel. It would be better for an impatient or careless driver to hit the EMS unit than the rescuers on the roadway. Once other emergency providers arrive, the EMS unit can be moved to a better location (as previously discussed).

• *Try to create one "non-traffic" lane next to the area where emergency crews are working.* When logistics allow, this will provide more of a buffer zone. Whether this is possible depends on such factors as the size of the

Figures 7-2A and 7-2B: From the approaching driver's point of view, cones tend to be more visible, as shown in these two photos.

incident, the number of lanes in the road, the location of the crash in terms of hills or curves, the weather, visibility and time of day. It is better to completely close a road for a short time than to risk the safety of responders simply to facilitate traffic flow.

• *Wear protective gear with reflective tape or trim.* At a roadway scene, put on protective gear with plenty of reflective tape or trim to promote visibility, This is particularly important at night and for those who walk away from the scene to direct traffic or place warning devices.

Proper use of warning devices

The purpose of warning devices is to alert motorists to a problem in the roadway in time to allow them to slow down or stop. Elements that affect stopping include the driver's reaction time and rate of travel, Warning devices may include the emergency lights on the EMS unit, reflectorized cones or flares. Although space to store them may be limited on EMS units, reflectorized cones are safer and more effective as warning devices than flares, especially in daytime (see Figures 7-2A and 7-2B).

The primary warning devices at most scenes are the warning lights on the emergency vehicle. Leave flashing and revolving lights on when working at a roadway EMS scene to warn other drivers. Headlights, however, can minimize or cancel the effects of emergency lighting for drivers coming from the opposite direction. Turn them off (unless they are needed to illuminate the scene>.

When possible, set out additional warning devices. There are three factors to consider. First, how visible is the crash site to oncoming drivers? If it is hidden by a curve, a hill or weather such as rain or snow, the importance of additional warning devices increases. Second, how long will it be until assistance will arrive? If police are only a minute away, the task can probably wait. Finally, will the EMS unit be on the scene long enough to make the effort worthwhile? When injuries are critical and extrication is straightforward and quick, the goal is to load and begin transport in ten minutes or less.

Place warning devices far enough away to give oncoming drivers time to stop. This distance is measured from the edge of the danger zone. Remember that the diameter of this zone increases with such hazards as downed electrical wires, fire and hazardous materials (see Figure 7-3). On roads with posted speed limits of 50 mph, the person putting out warning devices may be up to 225 feet from the crash. The figure shows distances for warning devices as they relate to passenger cars; roadways heavily traveled by trucks require extended placement to provide additional stopping time.

Whoever is placing warning devices must be acutely aware of traffic, since walking away from the crash site increases rescuer vulnerability. Because drivers may be distracted, EMS personnel should never turn their backs to moving traffic. Also, care should be taken to avoid placing flares in or near leaking flammables.

Principles of Disentanglement

A team of EMS personnel was dispatched to handle an incident on a roadway, Once in the area, they saw nothing-until they noticed a crowd of people on an overpass pointing to the steep embankment. There, about 60 feet up, was a minivan on its roof, prevented from falling to the pavement only by thick bushes. It had become airborne and jumped the 6-foot chain link fence at the top of the embankment; its unconscious diabetic driver was jammed into the pedals. Well-meaning bystanders had tied the van to the collapsed fence with a length of clothesline. As the paramedics arrived, they saw the vehicle sway as a first responder climbed over the top to reach the driver's window.

Fortunately, in this urban situation, proper rescue equipment was available almost immediately, and the vehicle was soon secured from falling further down the embankment. Entry prior to proper stabilization had been foolhardy; the first responders were lucky,

A crashed vehicle should always be stabilized-whether it is found on its wheels, side or roof, stacked up on another vehicle or hanging off the edge of an embankment. The cardinal rule is not to create additional victims. All that may be needed is to chock the wheels of upright vehicles. Securing the proper size and type of rope lines or cables to something solid (a tree, a fire hydrant or the frame of another properly braced vehicle) may be appropriate. Many EMS personnel have crossover training in disentanglement and extrication and carry the proper equipment on their EMS units. Others do not. Either way, proper stabilization helps ensure that victims will avoid additional injury; for them, even a slight rocking or bouncing of the vehicle could translate a spinal injury to permanent paralysis. Stabililization also minimizes the chances of injury to rescuers.

Never work on the downhill side of a crashed vehicle. If a vehicle is on its roof, the roof pillars may collapse at any time. Insert materials such as rescue airbags into spots that need more padding. Be sure no one, including the patient, will be trapped by an inflating rescue airbag.

Unreleased airbags, also known as supplemental restraint systems

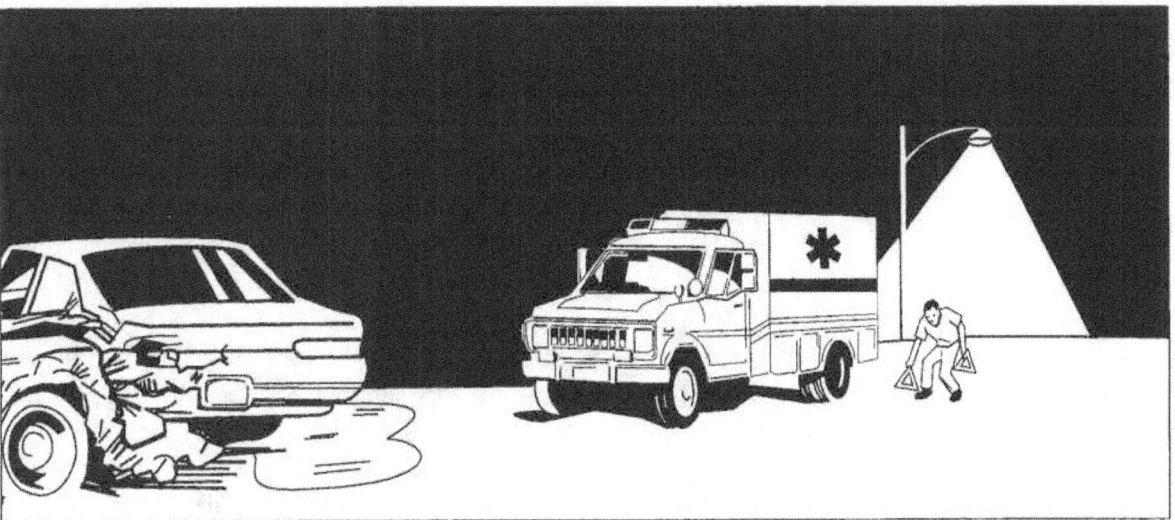

Figure 7-3: On-coming drivers must be given warning early enough to have time to stop. This figure demonstrates appropriate placement distances for traffic warning devices.

(SRS).[3] As of the 1993 model year, most new vehicles are equipped with airbags, which are designed to protect occupants in head-on (or near-head-on) collisions. Some have them only on the drivers side, but others are available for both front seats.

There are times airbags will *not* deploy, either because a crash was not severe enough or because it did not involve front-end forces. There are different triggering systems, but each generates pure nitrogen gas almost instantly, which deploys and fills the airbag in less than the blink of an eye. The time from initial impact until the bag deflates is less than the time it takes to sneeze.

A deployed airbag poses one minor safety hazard: the fine, chalky powder may slightly irritate the skin, nose and eyes. Flushing is recommended whenever skin contacts the powder; prevention for the responder is through use of protective eyewear and gloves.

Inadvertent deployment of an undeployed airbag during the rescue phase would be very rare, and the risk of injury is small. However, "not only could accidental inflation prove harmful or fatal for medical or rescue personnel working inside the vehicle, but a sudden jolt from the inflated airbag striking the trapped or injured patient could likewise have tragic results"[4]—such as additional or exacerbated injuries. The EMS providers most reliable protection against inadvertent deployment is to avoid working in the undeployed airbag's path. Another safety precaution is to disconnect the vehicle's battery; it is safest to disconnect the negative side first. Although the "drain time" to eliminate the chance of electrical charging of the SRS unit may be as brief as eight seconds, General Motors technicians are advised to wait ten minutes. EMS personnel may not have that luxury, so the sooner the battery is disconnected, the better.

Hydraulic bumpers. Energy-absorbing bumpers have been in use since the mid-1970s. Stand clear of them. The implicit hazard is that these piston-loaded bumpers can do more than return bumpers to their original positions. It is possible for compressed bumpers to mechanically release during the rescue process. The bumper's strike zone would be just a few inches if the bumper does not detach from the vehicle, but if it detaches, the strike zone may be considerably further. Minimize this potential problem by securing unreleased bumpers to the vehicle's undercarriage with a chain. A highly unusual scenario might be that "the mechanical impact [of a crash] itself may be strong enough to fracture the mounting bracket assembly, mechanically disconnecting the bumper from the piston tube. The heat from the ensuing fuel spill and fire can launch a piston tube assembly. The tube, or portions of it, can become an unguided missile with a potential travel distance of approximately 300 feet, the length of a football field."[5]

Hydraulic pistons. If it becomes necessary to cut off the rear door of a van or hatchback, those with hydraulic pistons can suddenly pop. Stand clear.

Because auto crashes are usually multiagency responses, good teamwork is important. The Incident Management System is helpful for coordi-

by Eric S. Lamar

It is unfortunate that death and serious injury are a common reality for firefighters and paramedics. It is a particularly disturbing fact that many firefighters' first experience with a line-of-duty fatality will not be in a fire situation, but rather at a highway incident. A general increase in highway speed, a rise in the number of substance-abusing drivers and the proliferation of hazardousmaterial-related accidents all have contributed to the number of serious highway incidents. The ten rules listed below should help to increase firefighter awareness of the particular dangers of highway operations and how to survive in what is becoming an increasingly hazardous area. Remember: firefighter protection is our responsibility.

P roceed slowly, using your vehicle as a safety shield for personnel.

R emember: passing motorists are watching you, not where they are going.

O bserve types of vehicles, placards, condition of containers and fire hazards.

T ake the time to stabilize all vehicles before beginning operations.

E vacuate as necessary to ensure the safety of public and fire department personnel.

C all for additional assistance early. Overtaxed personnel become a safety liability.

T reat the incident scene with great caution.
Wear full protective clothing and use all warning devices.

I n the interest of safety, appoint a safety officer/spotter and dedicate adequate resources for fire protection.

O nce size-up is complete, close the road when necessary to protect personnel.

N ever let your guard down in the take-up phase.

nating different emergency responders with different tasks. For example, if someone from one agency is in charge of disentanglement and someone from another agency is coordinating emergency medical care, the two must work together to avoid compromising patient care and to promote the safety of everyone involved.

Approaching a stopped vehicle

EMS units are sometimes dispatched to roadway incidents involving a single vehicle when no crash has occurred. The EMD reports "someone slumped behind the wheel" or a similar description. Sometimes, the person is simply asleep at the wheel. At other times, there is a true medical emergency, And rarely, people play possum and mean harm. The EMS provider has no way of knowing which scenario is true until after reaching

Figure 7-4: In some situations, such as drivers who are slumped over the wheel, approaching from behind is prudent until the circumstances are clear. Create a "corridor" that will protect the approaching investigator from other traffic by positioning the EMS vehicle as shown.

a relatively vulnerable position.

The element of surprise is a good tool in these cases. However, surprise is a two-edged sword. From one point of view, the noisy or obvious EMS provider provides ample opportunity for someone who means harm to prepare the attack. On the other hand, the individual who is actually just sleeping may respond violently out of surprise if the EMS provider sneaks up too quietly. Judgment and awareness are essential.

The suggestions in this section are in no way meant to create the impression that EMS personnel should be taking on police functions. When law enforcement personnel are available, they should be the ones to confirm the safety of the situation. Since EMS personnel are often left to their own resources, these suggestions are intended to provide safe methods for assessing potentially dangerous roadway scenes.

If law enforcement is not on the scene, approach stopped vehicles from

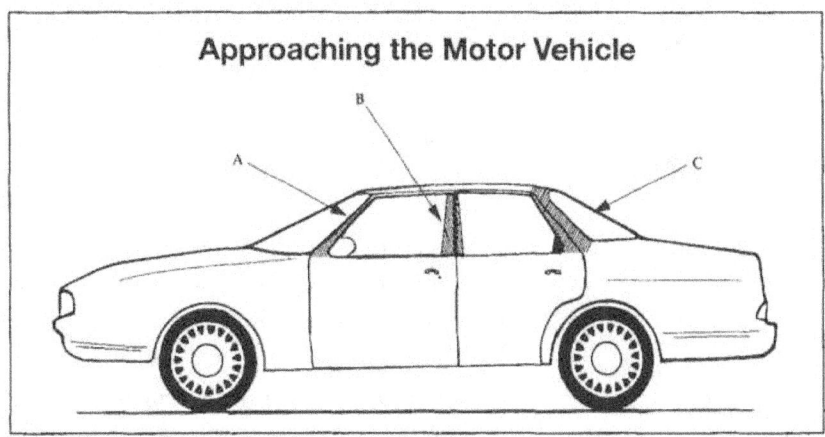

Figure 7-5: Never get forward of column B when investigating a situation. It is difficult for someone with bad intentions to cause harm accurately when shooting or striking over his shoulder.

behind. Park no closer than 15 feet away, and park at an angle that will facilitate rapid departure.[6] If the person investigating the parked car will be exposed to moving traffic, the EMS unit can be slightly offset to provide a channel of relative protection (see Figure 7-4). The driver should stay behind the wheel to allow for a quick getaway if needed.

First, try eliciting a response from the patient by using the PA system. If that does not work, the EMS provider approaching the stopped vehicle should walk behind the EMS unit (rather than between the vehicles) to avoid being silhouetted by the headlights at night and to avoid being crushed between the stopped vehicle and the EMS unit should either vehicle move suddenly, Upon reaching the EMS vehicle driver's door, check with the driver whether anything about the scene changed while walking behind the vehicle.

Leave the jump kit next to the EMS unit until after assessing the situation. Move quietly and quickly to the rear of the stopped vehicle. Check to be sure the trunk is closed; if it is unlatched, slam it shut without opening it. Some law enforcement officers have been fired upon by fugitives hiding in the trunk. Throughout the approach, watch the inside of the vehicle for signs of trouble. For example, if the side mirror is moving, or the occupant seems to be shifting suspiciously (like reaching under the seat for something), abort the approach, return to the rescue unit and call for immediate law enforcement backup.

Use the protection of the passenger vehicle itself. Most cars have "A," "B" and "C" columns (see Figure 7-5). Pause briefly at column C to check whether there is anything of concern in the back seat; then move to column B to observe what is happening in the front seat. Try to see occupants' hands, and avoid reaching into the vehicle, where it is easy to be grabbed and held. Never pass column B; doing so positions the EMS provider where it would be easy to be shot or knocked off balance if the door was opened quickly.

The person investigating the vehicle should loudly identify himself as representing emergency medical services to help promote the helpful purpose of his presence. If breaking into the vehicle becomes justified, use a different window (to protect the patient from glass fragments). One EMS provider should remain by column B to watch the patient-who may yet wake up and respond violently to the commotion.

In the event of a violent response to initial attempts at assessment, good positioning can be lifesaving. It is difficult for the person in the vehicle to shoot accurately over his shoulder at someone standing by column B. Keep something soft (such as a pad of paper) in the left hand; this can be used as a distraction by throwing it into the assailant's face, perhaps buying time for a safe retreat to the EMS unit.

Although approaching a stopped vehicle for EMS purposes has a different purpose than it tends to for law enforcement purposes, the distinctions may not be apparent to the occupant. Avoid complacency, and acknowledge the inherent hazards of this sort of encounter. Be careful.

Other Outdoor Considerations

The activities of arriving, approaching the patient and initiating interaction with people at the scene all have the potential to place EMS personnel in positions of vulnerability. Many EMS personnel, when approaching a situation, walk along existing pathways or straight to the subject. This is what those awaiting help (and those intending harm) expect. A good precautionary routine is to walk separately. This allows for two viewpoints of the scene; one partner may be able to see something the other cannot. Walking separately also creates two targets; if someone is intending to harm EMS responders, the chance of both being hit is thus lessened. When possible, walk along unconventional pathways when approaching the scene, thereby operating outside the expectations of others. That is, walk across the lawn instead of up the front walk, or go around the vehicle parked in the driveway instead of past it the easy way.

Although hostile weapons fire is rare, one should acknowledge the chance it could happen each time one is in an area that could be hit. This area is what police refer to as the "kill zone."[7] Until the scene is fully assessed, one never knows if a kill zone situation is in effect. Awareness of the kill zone must be especially acute at outdoor settings, since EMS personnel are often at the center of widespread attention (see Figure 7-6). A problem may even be launched from behind: One paramedic was assisting a homeless man one night when he was attacked from behind with a small pocketknife by an alcohol-impaired bystander. The paramedic was wearing body armor, which prevented serious injury.

In another case, the police called paramedics for an officer down at a place where crowds of angry people had been gathering frequently. The responding paramedics were protected by a ring of police officers standing in a circle, facing the crowd; inside the circle was the injured police officer. That system was very effective in protecting the medics.

Cover and Concealment

The consistent use of cover or concealment opportunities is an important

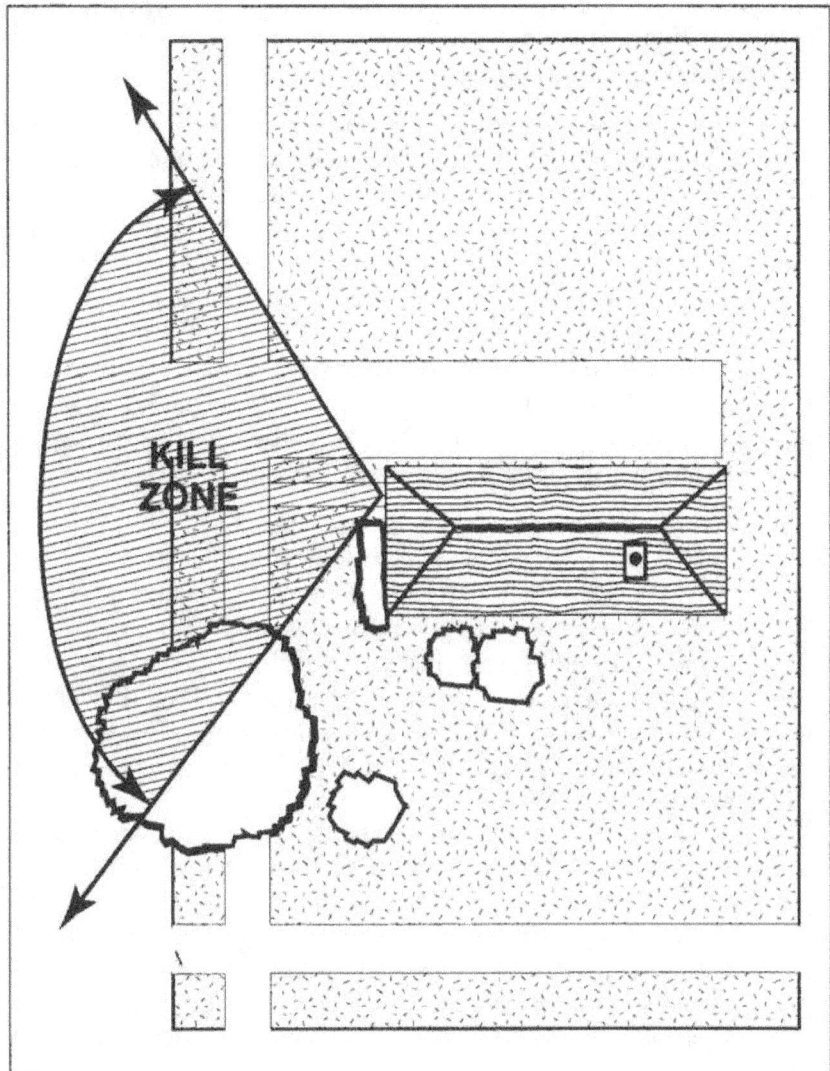

Figure 7-6: Always be aware that when entering the kill zone whether a hazard exists or not may not be known until it's too late.

prehospital tactic. They may be the best chance for survival in situations that are impossible to judge ahead of time. (Remember, law enforcement should precede EMS personnel to scenes known to be dangerous.)

The difference between cover and concealment is as follows:

• *Cover.* An object large enough to hide behind and strong enough to stop a bullet. Good examples include a solid tree or telephone pole, brick or stone walls, steel garbage dumpsters (the fuller the better). The door of the EMS unit or a wooden fence may both seem substantial but usually cannot stop a bullet, so they cannot be considered adequate cover. Remember, the best cover in situations that appear hostile is to stay in a safe location out of the way while law enforcement personnel secure the scene.

• *Concealment.* A place that will hide a person from view, but which may not be capable of stopping a bullet. The door of the EMS unit and the wooden fence previously mentioned would be suitable for concealment. A large bush, or even shadows, are also appropriate. It is important to be

Emergency Incident Rehabilitation

The circumstances of some emergency incidents make establishment of a rehabilitation area essential. It is recommended that emergency personnel work no more than 45 minutes between mandatory lo-minute (minimum) rest periods. There are certain characteristics to seek when selecting a rehabilitation area. It should provide appropriate protection from the prevailing environment. It is inappropriate to send personnel from a hot environment directly into air conditioning. Shade and fans may be more appropriate. The location should be beyond the sights and sounds of the incident, to allow safe removal of personnel protective equipment and to allow for mental rest. It should be free of exhaust fumes, large enough to accommodate multiple crews, easily accessible to EMS units and nearby enough to facilitate prompt re-entry to the incident upon release by the Rehab Officer.

The Rehabilitation Area should provide:

Medical evaluation, treatment and monitoring. Vital signs and general medical assessment should be done by qualified personnel (minimum BLS-level training). Heart rate should be measured and documented as soon as a person enters the Rehab area. If it exceeds 110 beats per minute, temperature should also be taken and documented. Anyone with an oral temperature in excess of 100.6 F should not be allowed to wear protective equipment. If the body temperature is lower than 100.6 F but heart rate remains above 110, rehabilitation time should be increased. The EMS staff should also aggressively assess for other potential medical problems,

Food and fluid. This is essential when an operation lasts more than three hours. A high level of hydration is critical for prevention of heat injury, The goal is to consume one quart of water per hour during strenuous activity. The recommended hydration solution is a 50/50 mixture of water and "activity beverage" such as those that are commercially available. Carbonated and caffeinated beverages, as well as alcohol, should be avoided. The food provided should be easy to digest and low in fat, refined sugar and salt. Stews, broths and soups are easier to digest than sandwiches or fast food. Excellent sources of quick energy replacement are foods such as apples, oranges and bananas.

Other. Depending on the circumstances of the particular incident, the Rehab Area may need to be stocked with dry clothing, heaters, blankets and towels, tarps (for shade or ground cover) and other supplies and equipment. It should be readily identifiable and demarcated with traffic cones and fire-line tape if it is located in an open area.

All personnel involved in a large emergency operation have the responsibility to monitor their own well-being. No one should be allowed to work past a certain time period without a rehabilitation break.

Exerpts from Emergency Incident Rehabilitation, FA-114, USFA, July 1992

sure that the structure is reliably concealing; for example, would the hiding spot be illuminated by the headlights of an approaching vehicle?

Thus, a second valid reason to walk along unexpected pathways is that they provide the most opportunities for cover and concealment. EMS personnel should continuously identify the nearest or best places that could serve as cover or concealment in case the need arises suddenly

Environmental Hazards

Rain, snow, ice, bright sunlight, extreme temperatures, humidity-each has an impact on an outdoor prehospital scene. Prehospital personnel are continuously thrust into the elements, yet they are no more immune to the elements than other humans. One paramedic slipped on the ice and fell at a scene, fracturing his leg. A second EMS unit was needed to transport him to the hospital. Another EMS crew was subjected to an intense downpour while getting a patient from the outdoor scene to the EMS unit. Visibility was poor, and EMS crews were exposed to moving traffic. Later, their health was threatened since there was no time between calls to change into dry clothes.

The environment is a meaningful physical stressor. Extremes and seasonal changes can diminish resistance to disease, and they can decrease emotional tolerance for the needs of others, sometimes with dangerous results. Those awaiting care may also be so affected by the elements that they are more irritable than they might normally be. Be aware how the environment might be affecting people and their moods,

Heat stress and cold exposure are significant concerns during labor-intensive operations or long-duration incidents and training exercises. The Incident Commander is responsible for ensuring that a Rehabilitation Area is established and maintained by the assigned Rehab Officer at such scenes. The purpose of Emergency Incident Rehabilitation is "to ensure that the physical and mental condition of members operating at the scene of an emergency or a training exercise does not deteriorate to a point that affects the safety of each member or that jeopardizes the safety and integrity of the operation."[8] (see Sidebar 2)

There are other outdoor environmental hazards. For example, an EMS crew was in the city park one night, helping a person who was lying on the grass, when the automatic sprinklers came on. Although this incident is amusing to describe afterwards, there is nothing funny about being hit at close range by streams of pressurized water. In the past, EMS personnel have felt compelled to help patients despite overhanging debris or exposed situations where hasty retreat is not possible. In an ideal world, the personnel who perform extrication or who secure the scene would bring such patients to a safe location for the safety of EMS personnel; at this point, the nature of EMS is that it is still sometimes risky. There will always be borderline circumstances that demand good judgment and the best possible safety measures in spite of imperfect settings.

Large Pre-Planned Events

A large event such as an outdoor concert poses the challenges normally associated with crowd control. Planning should be closely coordinated with other local emergency agencies. Pertinent issues include how access to patients will be ensured, with safe escort procedures being established for EMS personnel. The nature of a crowd varies according to the cause for gathering (a political candidate's speech, a heavy-metal concert, a celebration of Earth Day). Pre-planning should include briefing public safety personnel about safe zones, crowd control planning and whether additional security forces will be present. Cross-training with other public safety groups can promote familiarity with and confidence in each others' skills.

Summary

There are many outdoor hazards-some environmental, some interpersonal. Avoid unnecessary harm, and use every resource available for protection. Even small incidents that can cause harm are worth avoiding. Roadway operations are particularly dangerous, resulting in deaths among emergency workers every year. Many could be avoided with increased attention to safety A mental attitude of self-preservation, not complacency, is vital. The element of teamwork is important, so that each emergency responder at a roadway operation is aware of both the medical demands and the powerfully dangerous environment surrounding everyone.

References/Endnotes

1. Harvey D. Grant, et al, Emergency Care, 4th edition (Englewood Cliffs, NJ: Brady, 1986), p.497.
2. Grant, Emergency Care, p. 495.
3. National Highway Traffic Safety Administration brochure, "Emergency Rescue Guidelines for Air Bag- Equipped Cars" (DOT HS 807-579, Rev. August 1990.
4. Ronald E. Moore, "Passive Restraints and Safety," Rescue, May/June 1989, p.42.
5. Ronald E. Moore, "New Extrication Hazards: Surviving the New Auto Technology," JEMS, October 1985, p.30.
6. Dennis R. Krebs, et al, When Violence Erupts: A Survival Guide for Emergency Responders, p. 13.
7. RJ Adams, TM McTernan and C Remsberg, Street Survival: Tactics for Armed Encounters (Chicago: Calibre Press, 1980), p.49.
8. USFA, FA-114: Emergency Incident Rehabilitation (July 1992).

CHAPTER 8

Indoor Operations

Chapter Overview: There are numerous strategies related to safe indoor EMS operations. These include gaining entry safely, being aware of structural hazards (such as unsafe stairways) and positioning oneself safely during and after gaining access to the patient. Each is addressed in this chapter.

The mission of EMS-to care for sick and injured people-almost always means people are grateful for the assistance and mean no harm. The operative word, though, is "almost." The EMS provider must never forget that it takes just one moment to wipe out years of safe practices. Proper alertness for the possibilities for harm is constantly essential.

Once indoors, the EMS provider is on unfamiliar ground. Disadvantages include the fact that people at the scene have greater familiarity with the building, the structure itself may be unsound, and the EMS provider is frequently placed in ambush-prone positions.

Gaining Entry

Never go alone into a structure, especially large buildings with multiple floors. Even when escorted by a seemingly congenial stranger, the EMS provider who enters without colleagues is at the mercy of whatever or whoever is inside. Other emergency responders may be unfamiliar with the structure and may not know how to find that person quickly should a harmful situation arise. When emergency assistance arrives in sequential waves, as happens in tiered-response EMS systems, it is helpful to enlist an able bystander to provide directions for arriving crews. This avoids unnecessary depletion of personnel, who then can stay together.

Always carry a portable radio. Even if there is a telephone available to call for help, a radio is quicker and more direct. Also carry a good, reliable flashlight, even in daytime. The patient's location may not be well lit; sometimes, a room has few light fixtures or lamps, and bulbs are low-wattage. Patients are sometimes in places that are dark, such as crawl spaces, basements, attics and closets. When carrying an illuminated flashlight, hold it to the side. (see Figures 8-1A and 8-1B) This changes the target in case someone with a weapon decides to shoot.

When entering dimly lit areas, EMS personnel may be at a disadvantage until their eyes adjust. It can help to close one eye a few seconds before entering. Using that eye upon entry means a slightly quicker transition to the darker surroundings. Be cautious about turning on lamps. One known booby trap is a live round of ammunition in a lamp socket. When the well-intentioned rescuer turns the lamp switch without checking the socket, the result is explosive instead of illuminating.

When knocking on doors, always stand to the side. This simple habit can be lifesaving; anecdotes about people shooting through doors are numerous. In one incident, the perpetrator shot an arrow through the door![1] Some are unaware that helpers are standing outside. One elderly woman shot through her door on the assumption that neighborhood hoodlums were harassing her again. Others do not care who it is; they just want to be left alone and shoot cold-bloodedly, on the assumption that anyone in uniform is more authority than they want on their premises. (Career criminals often spend their time in prison studying police tactics. Since police also stand to the side, EMS personnel should know that some people now know that authorities will stand at the area next to the door, and thus they aim there. Nonetheless, many more reports of shots through

Figures 8-1A and 8-1B: Holding a flashlight directly in front of oneself (photo A) provides a perfect target for someone with intent to cause harm. It is safer, perhaps, to hold the "target" to the side (photo B).

the door have been made.) Make it a consistent habit to stand well to the side (see Figures 8-2A, 8-2B and 8-2C.

In some cases, locks or activated security systems thwart access to the patient. Because of the legal and safety implications, gaining entry to a locked area is best left to law enforcement personnel. Call for assistance immediately. Whenever an EMS crew is involved with breaking into a locked area, even with the law enforcement assistance, there are certain risks. Flying pieces of door frame, locks and shattered glass are hazardous. Stand well back. When assisting someone properly trained in forced entry, wear gloves, eye protection and other appropriate protective gear.

Because some people create their own homemade security systems, even casually knocking on the door and then entering an unlocked dwelling may result in harm. Residents who go to the trouble of devising these sometimes elaborate systems have usually not considered that EMS responders may someday be the uninvited guests. Homemade burglar alarm systems may range from startling-but-benign (such as audible alarms) to lethal (such as shotguns cocked and aimed at the door, with the trigger pulled by a string attached to the door handle when someone tries to enter>. Such systems might be found anywhere, but are more likely in less affluent areas where people cannot afford professional installation of security systems.

Figures 8-2A, 8-2B and 8-2C: Never knock while standing in front of the door (photo A). Stand well to the side, either separately, as shown in photo B or on the same side (photo C) to minimize the chance of injury in the event of an attack.

When entering a public building-particularly a crowded, noisy place like a drinking establishment-it may be difficult to discern quickly where medical services are needed. Step away from the door right away, and stay near the wall while assessing the scene. If any hint arises that law enforcement assistance is needed, retreat and request help. Do not re-enter until it arrives. (If the call is known to involve interpersonal violence ahead of time, await law enforcement assistance before entering.) For safety and to facilitate medical care of the patient, it may be necessary to ask the bartender to turn down the music and to turn up the lights. Obviously, depart as soon as it is possible to do so without compromising patient care.

Indoor Hazards

Structures can contain numerous hazards. Stairways are always potentially dangerous. Carrying someone up or down stairs involves shifts in the balance of the load. Stair chairs are easier to balance than full-length stretchers. When helpers rush due to the intensity of working on a critically ill or injured patient, stairs are an especially dangerous obstacle, regardless of how sturdy they are.

If the soundness of a structure is questionable, examine stairs carefully. If they are rickety, the EMS provider should seek an alternative. One paramedic was on the lower end of the stretcher while carrying a patient down wooden stairs. Suddenly, the stairs broke under the weight, and the paramedic crashed through, with the combined weight of the litter and the patient on top. His back was permanently injured. He had to quit doing the job he loved. Alternatives might include seeking another stairway, exiting a window into a fire department or utility company aerial platform or ladder, or decreasing the burden on the stairs by belaying the stretcher with ropes handled by a crew on the landing (see Figures 8-3A and 8-3B).

In many cities, abandoned buildings are increasingly used for shelter by homeless people. Abandoned buildings may be structurally unsound. Fires set to cook food or provide warmth may go out of control. Debris

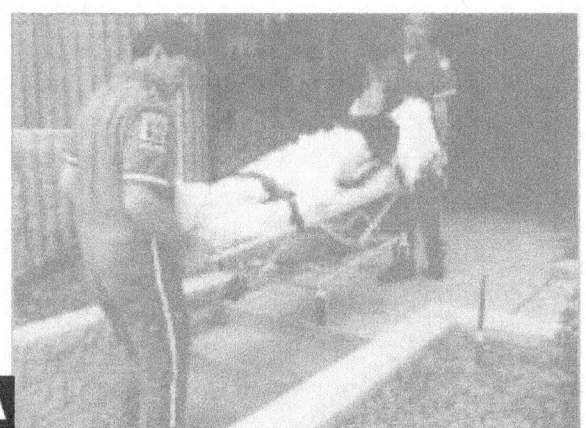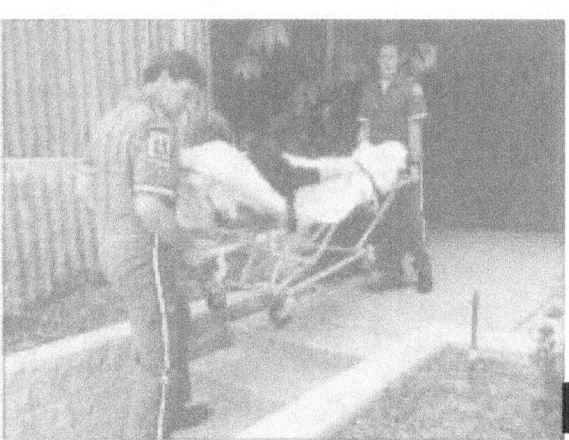

Figures 8-3A and 8-3B: Be aware of the best position for the patient when descending stairs. Some people advocate going feetfirst, so the patient can see (photo A); on steep stairs, however, this can make the patient feel as if she is tipping over, so descend headfirst (photo B). When possible, raise the head when descending headfirst. Communication with the patient is essential.

inside some buildings-and not just abandoned ones-can also pose a safety hazard. One EMT entered a room and found the patient on a filthy, unmade mattress that was lying directly on the floor. Intent on providing patient care, the EMT kneeled beside the mattress to check the patient's pulse-and badly cut her knee on glass mixed in the debris on the floor. From a hygienic angle, other emergency responders at one house found a carpet of feces and an overwhelming odor of urine. Dogs, cats and children were at play on the floor. Rotting food and dirty dishes clogged the kitchen. The situation was dangerously unsanitary.

When the police summon EMS personnel to a scene of violence, it is easy to assume that the scene is secure. In one case, paramedics were asked to take care of a shooting victim in a large facility, in which a former employee had shot several people. The EMTs assumed the perpetrator had been found and was in custody As it turned out, he was still at large-and any one of the people carrying guns and dressed in plainclothes they passed in the halls might have been him. They were lucky. Any time the status of scene control is uncertain, tactfully say something like, "Where is the person doing all this shooting? Is he in custody?" or "Is the scene fully secured?"

When a scene requires fire suppression or hazardous materials evaluation, the properly trained emergency providers must be summoned. EMS personnel may need to help set up a correctly designed receiving area for people as they exit indoor hazardous materials situations. It is important to know basic principles for what to do. Improper (or nonexistent) decontamination can have expensive, long-ranging consequences (see Chapter 9).

Animals are sometimes a greater hazard indoors because they know they are on their own home turf. Poorly trained guard dogs may attack without warning. Even normally placid animals may respond violently when their owners are in a medical crisis. Little dogs tend to hide under their owners' beds and bite ankles that come too close; these nuisances are usually easily removed, often by a relative. For more about animals, see Chapter 6.

At times, EMS personnel may be able to tell they are entering a potentially hazardous situation by the graffiti. Some instances of graffiti are just art, some done for the thrill of seeing one's name "in print." Other graffiti symbols are meant to define gang territorial boundaries, or as memorials to those killed in gang violence. Signs associated with satanic cults include the pentagram (5point star), the Nazi swastika, the iron cross, the cross of Nero (like the '60s peace symbol), the skateboard (as a symbol of anarchy), lightning bolts, 666 (or "FFF" since F is the sixth letter of the alphabet) and backward writing (including NATAS, REDRUM, LIVE and LIVED).[2]

Some people in fringe extremist groups-whether religious, drug trade, or political-are remorselessly violent. They may be heavily armed with sophisticated weapons. Many dislike any show of authority. If there is any question about what sort of people are being encountered, retreat until law enforcement secures the scene. Brief the arriving officers; sharing con-

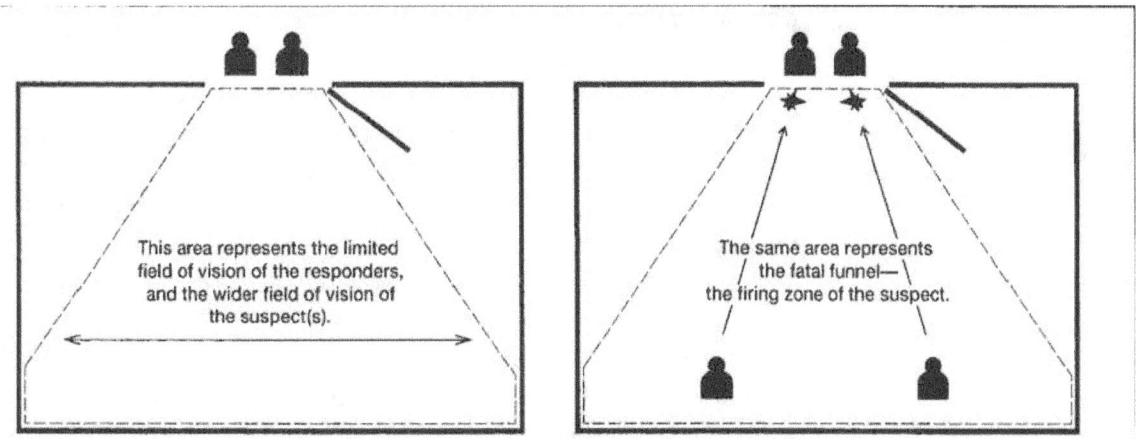

Figure 8-4: The fatal funnel places the person entering through the doorway directly in the line of fire from any angle inside the room. Be aware of this vulnerable moment, scan in all directions upon entering, and move away from the door quickly.

cern for each others' safety helps minimize the risk of harm to the different emergency response groups and demonstrates the compelling need for good interagency relations and dedication to teamwork.

Positioning for Safety

It is obviously dangerous to be trapped when a volatile situation erupts. The difference between being injured and escaping unharmed may depend on constant attention to positioning for safety.

Someone familiar with the building should be asked to lead EMS personnel to the patient. In some situations, a bystander or relative may point down a hallway and say, "She's down there, in the bedroom." Trusting EMS personnel who charge ahead with their mission do not realize the vulnerability of being caught in a narrow hallway between two strangers. Perhaps the person in the bedroom does not want the intrusion of EMS. Perhaps the person pointing the EMS responders down the hall knows trouble is likely to erupt but has not considered telling the well-intentioned EMS personnel the whole truth. Politely ask the on-scene person to lead the way.

In addition to hallways, particularly vulnerable indoor areas are stairwells, elevators and doorways. Each area makes the EMS provider easy to target. Entering a doorway is especially dangerous, since one is framed within what is known as the "fatal funnel." "From the point of view of one or more [people] in a room, all sensory concentration can be funneled down to just that small opening. It's your only way in, all they need to watch. Framed by it, you'd make an obvious target."[3] (see Figure 8-4)

When walking on stairs and down hallways toward as-yet-unassessed situations, six principles apply:

• *Don't look down.* Scan the surroundings, watching especially for signs that someone else wants to be hidden from view. Watch, too, for structural hazards such as missing boards and debris.

• *Walk seperately.* Bunching together creates an easy target. Walk on opposite sides of the hallway, or several steps apart. Look behind once in a while to evaluate what is happening to the rear.

- *Stay close to the wall.*
- *Don't chat.* Noise draws attention, provides a target and distracts the EMS personnel from the task of assessing a situation for safety. If communication is necessary, make it brief and to the point.
- *Glance around the comers first.* One crew startled a junkie who was shooting up when they rounded the corner at a landing. The junkie withdrew the needle from his leg and stabbed the EMT, injecting him with the drugs and HIV-contaminated blood.
- *Concentrate.* Arrival is a time of intense information-gathering for three reasons: patient care, communication and personal safety.

One very vulnerable indoors position is aboard an elevator. It is impossible to know what will be on the landing when the doors open. When a scene has involved law enforcement action, request police escort when riding an elevator. An alternative may be to get off the elevator one floor below the action and walk up, carrying a minimum of equipment.

In every instance, be aware of positioning in relation to the people at the scene. EMS personnel should never let the patient (or bystanders, when possible) get between them and the door, Always have an escape route in mind-and remember that the way out may not be the same way you came in. There may be secondary escape routes such as a back door, a porch door or even a window.

Once the patient has been contacted and assistance begun, a corner of the EMS provider's mind must remain in tune with safety. EMS personnel are most prone to tunnel vision during the phase when medical tasks are underway-evaluation, history-gathering, treatment, preparation for transport. Do not forget the surroundings. Things can still go wrong! Positioning for both patient care and safety becomes a challenge at times. Keep the doorway in peripheral view and assess newcomers to the scene. One EMS crew had their backs to the doors when someone new to the scene came in. They said later that they knew he was there but assumed it was a police officer. Instead, it was the patient's husband-the one who had beaten their young child earlier, and the reason the patient had taken the overdose. Had he intended to hurt the EMS personnel, he would have had two easy targets. Fortunately, he meant them no harm.

Getting out of an unsafe situation depends on the circumstances. It may work to use the excuse that both EMS partners need to go for a certain piece of equipment. Sometimes, how graceful a retreat may seem is less important than that it is done, such as when arriving paramedics were greeted at one door by a man holding a fistful of knives. Running was fully appropriate. They were grateful that day for being in good physical condition!

Summary

Indoor operations are usually safe for EMS personnel, but the potential for harm is high because the circumstances are restricted and unknown, An observant EMS provider should be able to avert harm while providing patient care. No one should ever feel foolish for deciding to retreat. Reasons for doing so may range from inability to gain safe access, to concerns about the safety of being in the structure itself, or because of violence-prone occupants.

References/Endnotes

1. Dennis R. Krebs, et al, When Violence Erupts, p. 74.

2. From notes taken at "Satanism & Cults" lecture by William Reisman, satanic cult specialist, March, 1991.

3. Ronald J. Adams, Street Survival, p.61.

CHAPTER 9

Hazardous Materials Situations

Chapter Overview: Proper hazardous materials response by EMS responders is an important element of this increasingly common type of call. Whether or not the EMS organization is the lead agency, all responders must have knowledge about hazardous materials terminology, basic safety strategies, the role of the EMS provider, specialized hazmat personal protective equipment and the decontamination process. This chapter addresses these principles.

Hazardous materials are one of the largest growing industrial realities of this generation. Four billion tons of hazardous materials are carried by air, surface, rail and water annually.[1] Proper manufacture, transport and disposal-often as hazardous waste-are difficult. They are used in thousands of ways, and they are everywhere.

For many EMS personnel, the idea of having to face a hazardous materials situation is unpleasant. An incident may involve a solitary patient with a minor chemical spill or entire communities, such as when a tank car full of toxic herbicide and pesticide crashed into the Upper Sacramento River in northern California in 1991. That incident affected the water supply of millions of people and wiped out wildlife in and around the world-class trout river for 40 miles.

Fortunately, the EMS provider has many resources for organizing both a scene involving hazardous materials and information about the substance(s) involved. In many places, specially trained hazardous materials teams with on-board references are prepared to respond quickly. In rural areas, where such teams may not be readily available, information about the substances involved is obtainable through national hazardous materials centers (such as CHEMTREC) or through regionalized poison control centers, which have on-staff toxicologists and abundant information.[2]

Once a hazardous materials incident is identified, there are three immediate basic tasks to complete. First is isolation of the event to prevent any further contamination. Next is identification of the material(s). The final task is recognition of the implicit dangers and need for decontamination. The importance of these basic tasks was highlighted vividly in a situation that occurred in Virginia in 1987. After an explosion at a computer manufacturing plant, the first fire company to arrive found some victims outside, with reports of a fire and about 20 more people inside. A triage area was established and the local hazmat team was en route, but patient transport was initiated prior to the team's arrival. The two hospitals were not notified that the explosion was the result of mixing two incompatible acids-and that the patients were soaked in one acid or the other, or both. Ambulance crews were overcome, and-in classic domino style-so were personnel in the emergency departments. Both EDs had to be closed and patients diverted elsewhere; one hospital had to be partially evacuated. None of the basic tasks listed above-identification, recognition of the danger and isolation-was done at the right stage.

Because each hazardous materials incident is unique, this chapter addresses only general principles, including terminology, basic safety strategies, role of the EMS provider as part of an intricate incident management system, personal protective clothing and equipment, and decontamination processes.

Basic terminology

A hazardous material is "any substance or material in a particular form or quantity which . . . may pose an unreasonable risk to health, safety, and property, or any substance or material in a quantity or form which may be harmful to humans, animals, crops, water systems, or other elements of the environment if accidentally released."[3] Hazardous materials are grouped into nine categories:

- explosives
- gases (compressed, liquified, dissolved)
- flammable liquids
- flammable solids
- oxidizers
- poisonous materials
- radioactive materials
- corrosive materials and
- other regulated materials (ORM)

When hazardous materials are released in an uncontrolled, unlicensed manner, the episode is known as a "hazardous materials incident." This may occur from a fixed site or during transport of the hazardous material. Those who use hazardous materials-in manufacturing, for example-are required by law to report the substances being used. This way, local emergency responders can know which specific substances might be involved in a hazmat incident. Transport vehicles such as railway cars and tractor-trailers are supposed to be placarded with basic information about the materials within. (see Figure 9-1)

A hazmat incident has several "control zones" to define sectors of protection. The "hot zone" is an area into which only those with appropriate protective clothing can enter. Typically, EMS personnel do not enter the hot zone because appropriate personal protective garments make most medical intervention very difficult. Rare exceptions would be for conducting triage or assisting with a trapped victim. The hot zone is sometimes also referred to as the "exclusion" or "restricted" area.

The "warm zone" is the control area immediately outside the hot zone. It is also referred to as the "decontamination", "contamination reduction" or "limited access" area. Only those properly trained and equipped to assist with decontamination are allowed. The inner and outer borders of the warm zone might be as far as 120 feet apart, depending on the scope of a particular hazmat operation.

The "cold zone" is also known as the "clean" or "support care" area. This is where the command post is located. Transport lines would form in this zone, to receive patients as they are passed outward.

The worst-case hazmat incident involves a situation where the amounts of the substances involved are immediately dangerous to life and health (IDLH). This is "the maximum level to which a healthy worker can be exposed for 30 minutes and escape without suffering irreversible health effects or escape-impairing symptoms."[4] In some cases, this is measured in parts per million-that is, in minute amounts. Another high-level classification is the extremely hazardous substance (EHS).[5]

One concern of hazmat experts is that untrained EMS personnel will be the first at a hazmat scene. This is one reason OSHA now requires completion of hazardous materials first responder training for all emergency re-

Hazardous Materials Placards

Some placards display symbols and/or are of certain colors. This is to show characteristics of the hazardous materials.

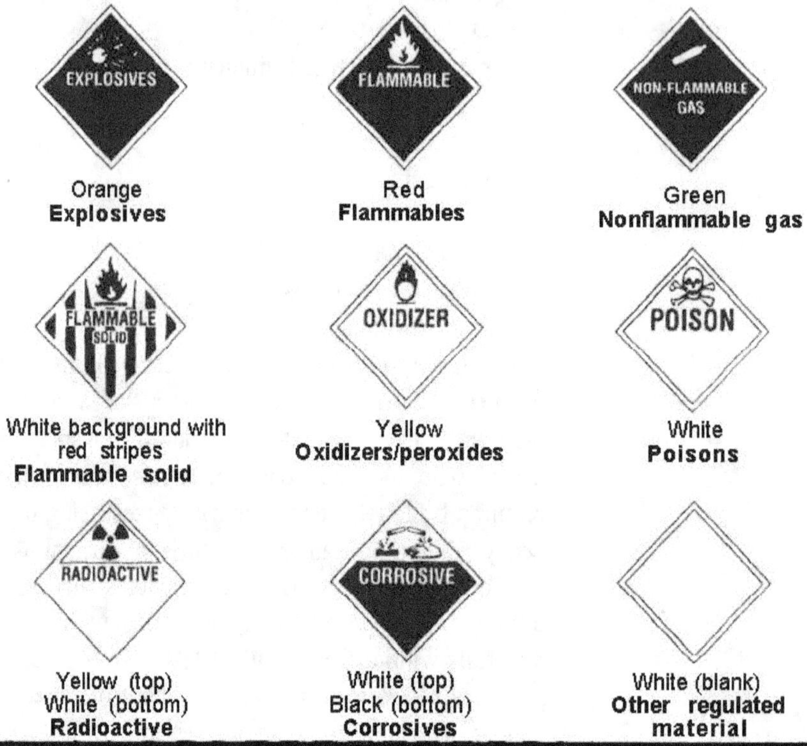

EXPLOSIVES
Orange
Explosives

FLAMMABLE
Red
Flammables

NON-FLAMMABLE GAS
Green
Nonflammable gas

FLAMMABLE SOLID
White background with red stripes
Flammable solid

OXIDIZER
Yellow
Oxidizers/peroxides

POISON
White
Poisons

RADIOACTIVE
Yellow (top)
White (bottom)
Radioactive

CORROSIVE
White (top)
Black (bottom)
Corrosives

White (blank)
Other regulated material

Placard symbols using the DOT system are indicated at right:

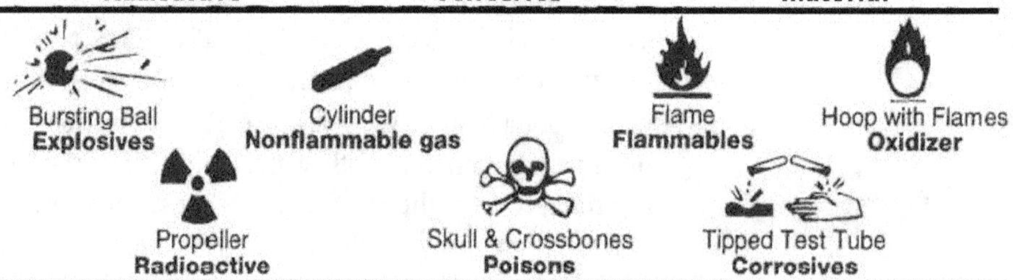

Bursting Ball
Explosives

Cylinder
Nonflammable gas

Flame
Flammables

Hoop with Flames
Oxidizer

Propeller
Radioactive

Skull & Crossbones
Poisons

Tipped Test Tube
Corrosives

Another identification which is widely used is the NFPA system of hazardous material identification, known as NFPA 704M. This system uses the same diamond shape as the DOT system, and subdivides the diamond into four more color-coded diamonds. Each is related to one of four parameters:

Blue = Health risk = blue
Red = Flammable material
Yellow = Reactive
White = Special characteristics or information (such as whether water can be used on the material

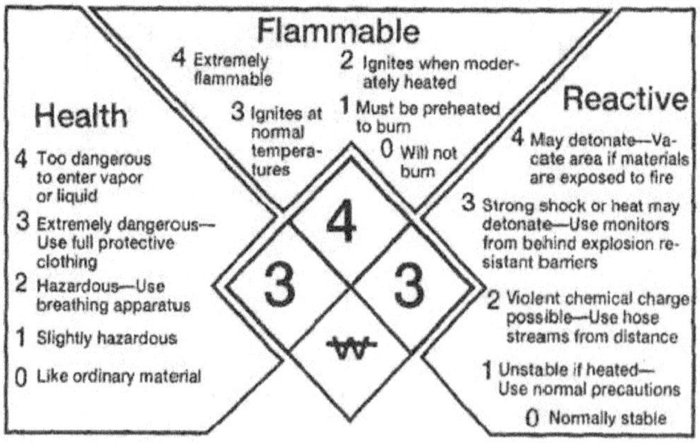

Flammable
4 Extremely flammable
2 Ignites when moderately heated
3 Ignites at normal temperatures
1 Must be preheated to burn
0 Will not burn

Health
4 Too dangerous to enter vapor or liquid
3 Extremely dangerous— Use full protective clothing
2 Hazardous—Use breathing apparatus
1 Slightly hazardous
0 Like ordinary material

Reactive
4 May detonate—Vacate area if materials are exposed to fire
3 Strong shock or heat may detonate—Use monitors from behind explosion resistant barriers
2 Violent chemical charge possible—Use hose streams from distance
1 Unstable if heated— Use normal precautions
0 Normally stable

4
3   3
W

In the center diamond-shaped areas contain numbers grading the level of risk in that area. The numbers range from 0 (no risk) to 4 (greatest risk). Symbols to indicate special information appear in the white area.

Figure 9-1: This figure shows a variety of placard designs. Trained hazardous materials teams know how to properly use the information they contain.

100

sponders, including EMS personnel. A hazmat incident can have a huge impact on a community, so good teamwork among local emergency agencies is essential. With a widely understood and accepted plan, a hazmat incident has the best chance of being handled smoothly, The local emergency planning committee (LEPC) is charged with forming a comprehensive local emergency response plan. No one expects a community to develop a plan overnight, but efforts to create a plan are necessary in case a large hazmat incident requires the coordinated efforts of local emergency personnel. For example, unless everyone concedes that a particular organization will be responsible for setting up the command post, several might vie for the task or worse yet, no one will assume the responsibility Either way, efficient scene operations will be compromised.

Regional hazmat planning is done through discussion and interagency compromise. The time to "fix" a problem is through the planning process, not during an actual event. EMS personnel are encouraged to take a proactive role in local hazmat planning.

Basic safety strategies

Basic safety strategy begins with each emergency responder respecting the importance of performing assigned duties without injury to self or others, property, or the environment. EMS personnel will first need to allow the hazardous materials team to properly assess the situation and set up for entry into the contaminated area. These things require time, so EMS personnel will need to do something unusual: wait. DO NOTHING. Defer to others for leadership and control of the situation. For the safety of everyone concerned, normal treatment processes have to be altered when dealing with contaminated patients.

Safety begins with the approach to the scene. Hazmat experts recommend that EMS units approach the scene uphill and upwind to prevent inadvertent contamination. This might be possible in some circumstances, when the caller knows exactly what has happened and the EMD knows the correct questions to ask. But sometimes EMS responders drive completely into the scene before discovering the true nature of the call. Should such an event occur, the on-scene crew must have the presence of mind to warn other responders and prevent too many people from driving into the hot zone. Others forewarned in this way should have the self-discipline to stay in their appropriate zones. Ideally, EMS units should be parked facing in a direction that allows rapid departure-that is, facing in a direction that allows for immediate forward motion.

Hazardous materials are identified by two primary systems. One system was developed by the US Department of Transportation (DOT), and a related system was adopted by the United Nations (UN). The second system of placard notification was established by the National Fire Protection Association (NFPA).

The DOT/UN system uses a diamond-shaped placard (on transport vehicles) or label (on other containers). Placards display the class code of the substance by way of a four-digit identification number within an 11-

inch diamond, which must be displayed on all four sides of the transport vehicle. These placards and labels are required on all transport shipments of the most dangerous hazards: radioactive materials, water-reactive flammable solids, explosives and poisons of certain especially high danger levels. They are also required to be posted when a shipment of hazardous materials considered slightly less dangerous contains more than about 1,000 pounds. Emergency vehicles should carry an Emergency Response Guidebook (ERG), provided to all emergency response agencies by the U.S. Department of Transportation. The ERG contains a list of the 4-digit codes and the chemicals they represent for rapid identification of the hazardous materials involved. The ERG also lists initial actions to take for incidents involving each chemical listed. All emergency units should also be equipped with binoculars to allow shipment identification from a safe distance.

Once a hazmat situation has been recognized, the responders must initiate the local system of response. The area must be quickly isolated and entry denied to non-essential and non-authorized individuals. This will prevent further persons from becoming contaminated. Initial responders can also begin establishing the hot, warm, and cold zones. Once in a zone, individual responders must stay there until proper decontamination procedures have been done. The best way to prepare for the chaos and uncertainty that initially surrounds setting up a hazmat response is to implement the local system in practice on a regular basis and to educate all who might help with mutual response how they would be expected to participate.

Role of the EMS provider

Participating in a hazardous materials incident is not the time to challenge system-wide plans. At a hazmat incident, the role of EMS personnel may be any of the following:

• Transporting decontaminated patients from the cold zone to an emergency facility

• Assisting in the cold zone with re-triage of patients as they are handed across from the warm zone and providing medical assistance prior to transport.

• Evacuating medical facilities that lie in an area of risk.

• Providing shelter support at locations where evacuees have been temporarily settled.

• Monitoring members of the hazmat team.

• Assisting with decontamination in the warm zone, if one has received prior training.

• Entering the hot zone to help with triage or to provide medical assistance for trapped victims. The ability to provide medical assistance-even the simple use of a stethoscope-is limited, given the restrictions of required personal protective equipment. EMS personnel trained both in hazardous materials incidents and the basics of triage are the only ones who should provide this service.

In the initial phases of setting up for a large-scale hazmat incident, extra EMS personnel may be needed for tasks such as assisting in the evacuation of downwind housing or offices. Because OSHA's document 29 CFR 1910.120 requires implementation of an incident management system for all hazardous materials incidents, EMS operations will probably be organized into a sector, division or group, with personnel reporting to the EMS Officer-who, in turn, reports to the Incident Commander. (For more information on Incident Management Systems see Chapter 6.)

Personal protective equipment

PPE is vital when dealing with hazardous materials incidents. It is sheer irresponsibility to walk in street shoes up to an overturned tanker, possibly stepping through some of the spilled material, to get a close look and to smell (and even taste) the material dripping out-but people have done it.

EMS personnel who will enter the warm or hot zones need to wear appropriate protective clothing and equipment. There are four basic levels of protection, with the highest being Level A and street clothing being considered Level D. (see Figure 9-2) NFPA standards address levels of clothing as well. A person wearing Level A equipment is theoretically protected for short exposures to the IDLH level and for prolonged work (such as clean-up operations) in contaminated areas identified as having a short-term exposure limit.

Air purification devices or self-contained breathing apparatus (SCBA) help minimize the greatest uncontrollable risk to emergency responders: inhalation exposure. Equipment such as respirators should be fit-tested before use. The wind may shift, or a contained event could suddenly destabilize and contaminate the air. Medical personnel should be trained how and when to use this equipment. Air purification devices, for example, should never be used where the ambient oxygen level may be below 19.5 percent. Proper storage will help insure proper function should this equipment be used. Equipment should be checked frequently

Because hazmat garments are so wide-ranging, EMS personnel are encouraged to know ahead of time what equipment will be available when the community-wide hazmat response is implemented, and how to use it should entry to the warm or hot zone become necessary, Cross-training with the local hazmat experts is the best insurance for a proper and safe response. For example, there are approximately 1.5 million different chemicals in everyday use in this society; because different chemicals require different types of protective gear, it is not practical for everyday EMS vehicles to carry chemical protective clothing. It should be provided by the hazmat team and may be stored somewhere other than an individual's EMS agency

EMS personnel whose roles are limited to patient transport and other cold zone contact should still be prepared to wear various levels of personal protective equipment, including gloves, face shields or safety goggles, masks and overboots.

EMS units used to transport patients from the scene of a hazardous materials incident must be properly prepared. Typically, patients will have been adequately decontaminated by the hazmat team. Two approaches to

Four Levels of
Personal Protective Equipment (PPE)

Level A

- Positive-pressure Self-Contained Breathing Apparatus
- Fully encapsulating chemical-resistant suit
- Double layer of chemical-resistant gloves
- Chemical-resistant boots
- Airtight seals between the suit, and the gloves and boots

Level B

. Positive-pressure Self-Contained Breathing Apparatus
- Chemical-resistant, long-sleeved suit
- Double layer of chemical-resistant gloves
- Chemical-resistant boots

Level C

- Full-face, air purification device (respirator)
- Chemical-resistant suit
- Chemical-resistant outer gloves
- Chemical-resistant boots

Level D

- Equipment does not provide specific respiratory or skin protection and usually consists of regular work clothes

Figure 9-2: There are four levels of Personal Protective Equipment (PPE). EMS providers who are operating strictly in an EMS role should not be needed to enter areas requiring high levels of PPE. Never attempt to use equipment for which appropriate training and education has not been received.

deal with this have been mentioned. Which procedure to use depends on local guidelines and the nature of the incident.

The first concept involves lining the vehicle in plastic. Unload unnecessary equipment and set aside equipment needed for direct patient care. Tape compartments closed with wide, heavy duty tape (such as duct tape). Tape heavy-gauge plastic sheets from ceiling to floor on all four sides of the patient compartment and to the ceiling as well. The floor should be covered with heavy-duty plastic. Only portable equipment will be used, and this equipment should be returned to the patient care area after sheeting is complete. It will need decontamination later. During transport, the EMS unit should be well-ventilated to minimize inhalation hazards[6] Mark the EMS unit as one prepared for transporting hazmat incident patients by placing wide strips of red duct tape in an X-shape over the Star of Life. EMS personnel should wear personal protective equipment (mask, gloves, gown, eye protection) and should promote air exchange inside the vehicle by opening windows or using the fan.

The second technique involves placing the patient in a "cocoon" (such as a body bag) with only his or her face exposed. This prevents contaminants from escaping. EMS personnel should wear appropriate personal protective equipment and use the vents and windows in the ambulance. There may be times, however, when a solitary victim is transported to be properly decontaminated at the medical facility. For instances when complete decontamination is not done at the scene, there should be a decontamination room at the medical facility with a special entry and pre-established arrival procedures that will protect the facility and the people inside. Coordination with local hospitals is critical; some facilities refuse to accept patients who are not fully decontaminated.

Decontamination processes

One of the most important principles of handling hazardous materials is to prevent further contamination. Decontamination must occur for all people and equipment involved. The EMS vehicle should go out of service after transport is complete until decontamination is insured. Inspection (and exams of personnel) should test for the specific substance(s) involved in the incident.

Even those who never went closer than the cold zone, but who were in contact with patients coming from the hot zone, need to be checked for, and cleared of residual contamination. In addition, all materials used for decontamination, such as the water used for washing and flushing the area, must be disposed of properly.

Proper decontamination depends on good planning, where the principles of hazmat control can be applied to the particulars of each unique instance. Advance planning helps ensure that proper and sufficient equipment will be available. An appropriate chain of command will exist, and mechanisms should be in place to protect the public and the environment from inadvertent secondary contamination. (see Figure 9-3)

There are four general types of decontamination:

• *Mechanical decontamination* involves brushing or wiping substances (such as dust or powders) off the victim. This is the first step of gross decontamination.

• *Dilution* is most often done with regular water, except when the hazardous material is water-reactive (such as sodium or lithium) or non-soluble (in which case, water with detergent added should be used).

• *Degradation* is chemical alteration and neutralization of the hazardous nature of a material. This process is used on equipment but is not practical for humans.

• *Isolation* of materials used for decontamination and residual hazardous materials is important. Collection systems for water runoff, for

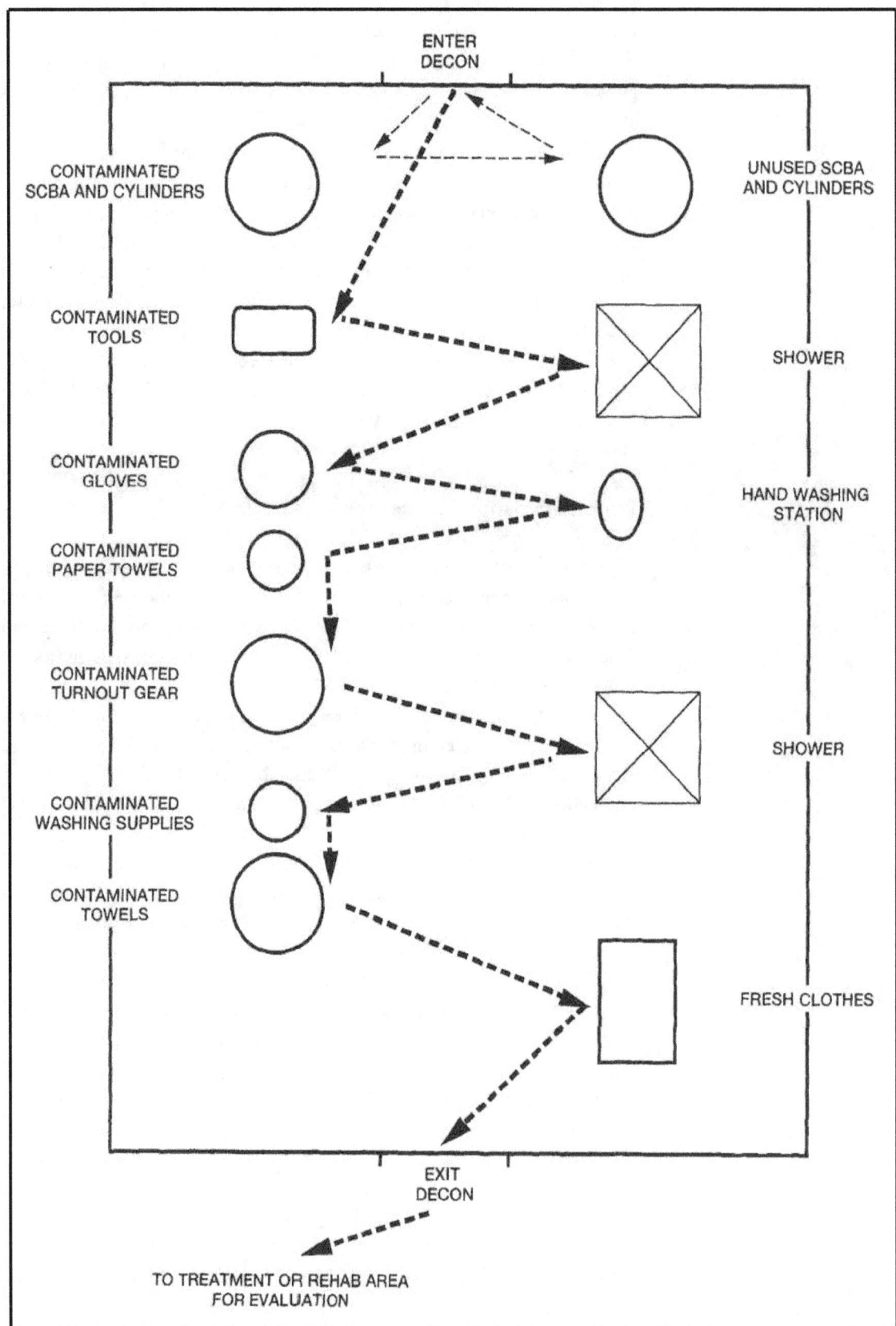

Figure 9-3: Proper decontamination is a multi-step process. Proper decontamination is vital for public safety.

> **Sidebar 1:**
> ## Resources for Information
>
> There are resources available for discovering what to do for contamination. They include:
>
> - **CHEMTREC** (800/424-9300), a private service for information about chemicals involved in transportation accidents
>
> - **ATSDR** (Agency for Toxic Substances and Disease Registry) at 404/488-4100, a 24-hour service that provides toxicologic information and hazmat incident guidance
>
> - **Centers for Disease Control and Prevention (CDC)** at 404/633-5313 (Georgia) for information about biologic and disease-related hazards
>
> - **The local poison control center.** Be sure to use one of the nearly 40 regional centers that cover all regions in the United States.
>
> - **National Pesticide Telecommunications Network** (NPTN) at 800/858-7378, a 24-hour service for information related to pesticide exposures and accidents
>
> - **Nuclear Regulatory Commission** (NRC) at 301/951-0550 with 24-hour assistance regarding radioactive materials

example, are very important; it does no good to flush a hazardous material into the city's drainage system.[7] (For information about what to do with a certain substance, see the resources listed in Sidebar 1.)

There are various philosophies about decontamination. The basic principle is a station-by-station approach that allows maximum control of the situation. The following nine-step procedure is an example

Station 1: Entry point and hot zone. A single entry point into the hot zone with a single exit to the next station. Mechanical decontamination of victims to the best extent possible is done here. Tools used in the hot zone are left in the tool drop area.

Station 2: Gross decontamination. Showering and scrubbing is done by people wearing appropriate personal protective equipment. Contaminated clothing is also rinsed off, then removed and isolated in bags. Stations 1 and 2 are in the hot zone.

Station 3: PPE removal. Personal protective equipment used by rescuers is removed, bagged and labeled for appropriate subsequent handling. If dressed in multiple layers, the rescuer should undress in stages, moving from one location to another within the station.

Station 4: SCBA removal. Removal and isolation of SCBA, or changeover of SCBAs if re-entry to the hot zone is necessary.

Station 5: Personal clothing removal. All items being worn (including jewelry and other personal items) are removed, bagged and labeled. Patients who were not undressed previously should be undressed here. Setting up this station requires discretion regarding the weather and privacy.

Station 6: Body washing. Thorough body washing is done, using detergent or soap and scrub brushes or sponges. Particular attention is given to nail beds, skin folds, groin and hair. Rinsing should be copious, with rinse water collected for proper disposal.

Station 7: Dry off. Towels or sheets must be provided for drying, and uncontaminated clothing supplied. Inexpensive, disposable garments are commercially available.

Station 8: Medical assessment. Few EMS personnel ever get closer to the incident than this point- highlighting the importance of knowing how to stand back and let others who are specifically trained and educated take charge of hazardous materials incidents. Medical triage, treatment and transport occur.

Station 9: Transport. Patients are loaded and transported to definitive care by properly prepared EMS units.[8]

Summary

The potential range and scope of a hazardous materials incident is daunting. The best approach is education, regional planning in cooperation with all the emergency agencies that would be involved and development of a specialized team. EMDs should have instant and accurate access to resources. Optimally, command officials should be known and trusted by the field providers. People interested in being visible but without a proper role (such as politicians) must be tactfully excluded.

Hazardous materials situations may occur in conjunction with other major disasters, such as earthquakes, tornadoes and hurricanes. Resources may already be stretched. Appropriate resources must be summoned anyway should a hazmat incident occur. The most important principle is interagency preparation and cooperation.

References/Endnotes

1. Jonathan Borak , Michael Callan, and William Abbott, Hazardous Materials Exposure (Englewood Cliffs, NJ: Brady, 1991), p. xi.

2. Kate Democoeur, "Poison Control Centers: They Can Ease Your Load,"JEMS, January 1986.

3. FEMA, "Glossary of Terms," GT 311: Hazardous Materials Contingency Planning Course (Revised), September, 1990.

4. FEMA, Hazardous Materials Contingency Planning Course (Revised), "Glossary of Terms," p. G. 13

5. US DOT, GT 311: Glossary of Terms: Hazardous Materials Contingency Planning Coure (Revised), September 1990, p.Gl2 (under "Hazards Identification")

6. Borak, Hazardous Materials Exposure, pp. 179-182.

7. Borak, Hazardous Material Exposure, pp. 145-6.

8. Borak, Hazardous Materials Exposure, pp. 149-152.

CHAPTER 10
Helicopter Safety

Chapter Overview: This chapter addresses the principles surrounding safe use of a helicopter, including choosing (and marking) a landing zone, providing appropriate crowd control and approaching a helicopter properly.

The size and load capacity of helicopters have evolved tremendously since the first commercial, hospital-based EMS flight program began in 1972 in Colorado. However, the principles of safety have remained essentially the same. Even though the presence of a helicopter adds drama and intensity to a prehospital scene, safety violations can be tragic and costly

Helicopter safety tends to be heavily stressed in areas where helicopters are used frequently, Ground crews must have sustained respect for this prehospital tool and not become complacent about it. An air of familiarity can be hazardous.

Basic principles of helicopter safety are as follows:

• Know when air transport is appropriate, and use helicopters only when properly indicated.

• Be sure a proper landing zone (LZ) is established. Know the methods used for daytime and nighttime landings.

• Be sure loose items near the LZ are secured before a helicopter makes its final approach to the scene.

• Have fire suppression available. Some helicopter services will not land if fire suppression is unavailable.

• Provide proper crowd control, both for casual bystanders and nonessential emergency personnel. Each must be kept well away from the landing zone and staging area.

• Be sure appropriate eye and head protective gear are worn.

• Use proper etiquette and technique when approaching the helicopter.

Air transportation is not a panacea. It is not appropriate to use a helicopter simply because the technology is available. Whenever a helicopter flies, there are certain inherent risks, such as crashing or being needed more urgently elsewhere. These risks must be outweighed by one or more of the three main benefits of using a helicopter:

First is when rapid transport over long distances is needed. Helicopters are an asset to rural EMS personnel; critically ill or injured patients often have access to large urban centers in a fraction of the time it would take to get there by ground transport.

Second is when terrain interferes with ground transportation. In some circumstances, a helicopter can airlift a patient over terrain that would otherwise require undue effort by ground crews; mountain rescue crews particularly appreciate helicopter capability (when a level LZ can be established). Ground transport may also be disrupted by earthquakes, heavy snowfall and other natural disasters.

Third is when advanced life support (ALS) is needed but unavailable from ground crews. The demands on providers to maintain medical expertise are heavy enough at the basic and intermediate level. When ALS-proficient helpers who maintain their skills through the volume possible in an urban area can respond rapidly in a helicopter, the patient may benefit.

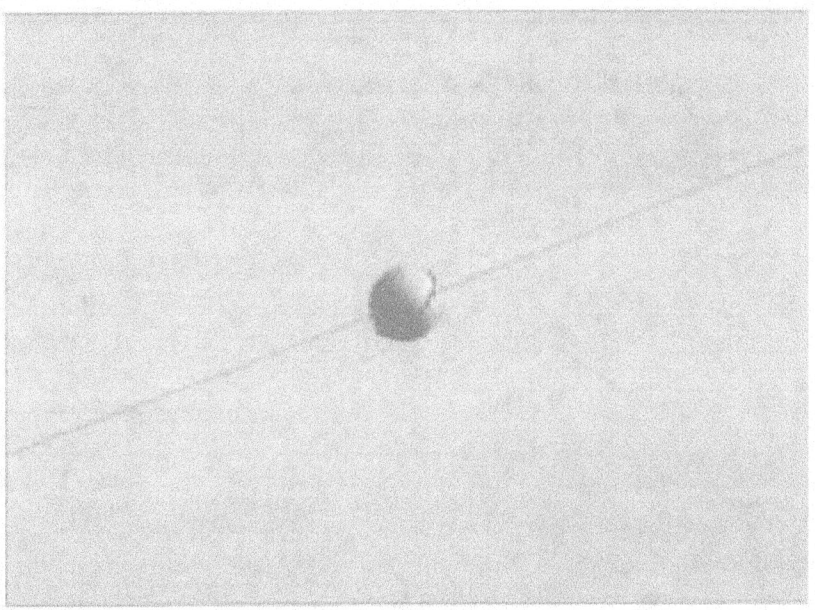

Figure 10-1 : The purpose of the globes hanging on the top wire is to notify helicopter pilots of the presence of wires, which are difficult to see from the air.

The Landing Zone

A proper landing zone (LZ) consists of an area large enough to accommodate the particular needs of the responding helicopter. A small helicopter (e.g. Jet Ranger or similar aircraft) may require a landing zone only 60 feet square; larger aircraft (e.g. Bell 412 or similar aircraft) may need up to 120 feet on each side of the landing zone. Size of the LZ can be estimated by pacing the distance; most people have approximately 3-foot strides, so 20 strides would constitute a 60-foot side for an LZ. The most vulnerable period for the helicopter is during takeoff and landing, when maximum stress is being placed on the engines, transmission, and rotors. Pilots generally prefer to take off and land at about a 45 degree angle, since vertical maneuvers and hovering also place maximum stress on the engine. This is why some pilots may prefer a rectangular landing zone. Local LZ guidelines should be known in advance; some agencies maintain preplan books that list known landing sites within their jurisdiction.

The LZ should be as level as possible. A pitch greater than about 8 to 10 degrees is risky. A sloping landing zone means the uphill rotor blades are closer to the ground, increasing the potential for harm should someone approach the ship from that side. It may also damage the helicopter if a rotor blade touches the ground or low-lying bushes.

The landing zone must be free of obstructions overhead, as well as along the angle of approach and departure. Electrical, telephone or guy wires are invisible to the pilot. (see Figure 10-1) Notify the pilot ahead of time of any potential airspace hazards (such as high towers) that are near the landing zone. Helicopter crash statistics illustrate the importance of this tactic; one report cited 15 crashes involving mechanical failure, 12 of which involved obstacles-all but one during approach or departure.[1]

Proper marking of a landing zone depends on the time of day and upon

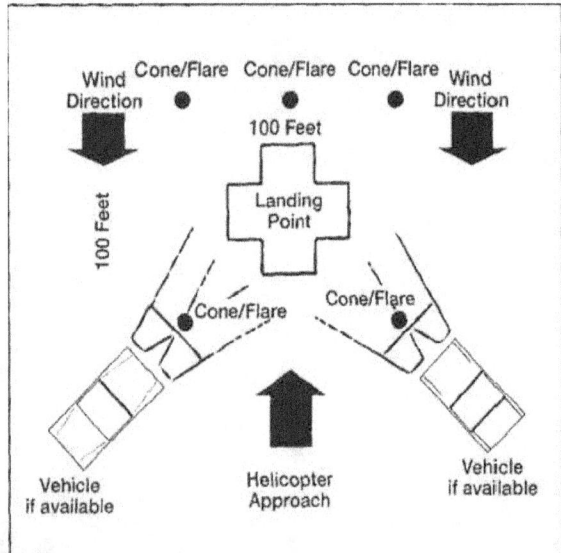
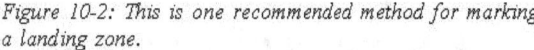
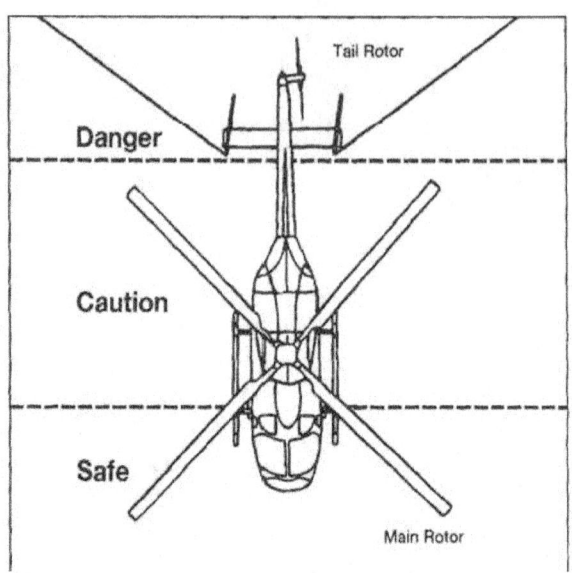

Figure 10-2: This is one recommended method for marking a landing zone.

Figure 10-3: Never approach a helicopter without the express permission of the pilot. Approach only from the safe zone, and never from outside the pilot's field of vision.

local pilot preference. In daytime, the four corners of the LZ should be marked with flares, cones, or approved landing zone strobe lights. Be sure marking devices are weighted so they are not blown out of position by the rotor wash. A fifth marker is placed at the center of the windward side of the LZ; its purpose is to show wind direction and speed. "Windward" means that side of the LZ from which the prevailing wind is blowing. (see Figures 10-2 and 10-3)

At night, a landing zone can be marked with vehicular headlights, if possible, or by other adequate light sources. However, it is paramount to protect the pilot's night vision. Never shine a light toward the helicopter, and avoid placement of lights in a way that they could blind the pilot at any point during the approach or takeoff. The pilot can lose night vision (essential to safe flying at night) for several minutes if just one flashbulb goes off the in his or her eyes, Flash photography must be avoided, and television lighting systems should not be allowed.

Local practice dictates whether someone should use arm signals to assist the pilot during landing. Most will not trust the person signaling unless prior training and a personal connection already exist.

When choosing a landing zone, look for a site near the patient, but not so close that rotor wash will be a problem. Parking lots are often good locations. Bushes, fences and any small obstacles should be marked with brightly colored surveyor's tape that is securely attached. Loose equipment or debris which could be blown into the air by rotor wash-such as the sheet over the patient or light backpacks-should be removed or secured to avoid their being sucked into the rotors. When possible, sandy or dusty terrain and areas with abundant loose, fresh snow should be avoided.

In some situations (for example, mountain rescues), it may not be possible for fire suppression equipment to be available when a helicopter is involved. However, when possible, fire equipment should be deployed and prepared to react should anything go wrong. Fortunately, fire service personnel are usually at these scenes in other capacities, as hazardous materials crews, emergency medical providers or extrication specialists.

Emergency Personnel and Bystanders

Proper crowd control is essential. The Incident Commander should delegate coordination of the LZ to another member of the emergency team. This person, usually the officer on the fire suppression unit, is responsible for preparing the scene for a helicopter arrival. Depending on the size of the scene, this person may also need to delegate certain tasks. In addition to establishing the LZ, good judgment must be used to keep bystanders who are not involved in emergency assistance well away from the landing zone. Some sources suggest at least 200 feet. Helicopters are captivating to many people, and some will try to get a closer look. Yet few bystanders appreciate the potential lethality of approaching a helicopter without the express permission of the pilot.

In addition, emergency personnel are sometimes even more enthralled about helicopters than bystanders and may be only slightly more educated about helicopter safety Non-essential emergency personnel should stand back at least 100 feet. No one should be allowed within the landing zone during arrival or departure of the helicopter. Once the aircraft has landed, only those with direct responsibility for tasks within the landing zone-such as loading the patient-should enter. When extra personnel are available, the person in charge of the LZ at each scene should post a lookout to prevent anyone from entering the LZ.

Helicopters can create winds greater than 60 miles per hour, inevitably throwing debris into the air. Every person within range of rotor wash should wear eye protection and head gear with fastened chin straps. Be sure the patient's face is well protected.

Approaching the Helicopter

The pilot is always in charge of who shall approach the helicopter. Depending on the circumstances, the pilot may elect to wait until the rotors have stopped turning before permitting anyone to come near; in more critical circumstances, it may be necessary to approach a helicopter while the rotors are still turning. When approaching a helicopter while rotors are turning, move with a slight crouch, particularly if unfamiliar with the equipment. Although there is plenty of headroom clearance on some helicopters, circumstances of rotor height may be altered by the terrain. Always approach and load a helicopter from the downhill side, and be sure IV poles, radio antennas and other tall objects will not hit the rotor blades.

The main rotor of most helicopters is usually between seven and 10 feet off the ground, but innate flexibility may cause it to dip toward the ground during deceleration. On sloping ground, the main rotor may be low enough to the ground to be hazardous.

The only safe approach is shown in Figure 10-3, within view of the pilot and with the pilot's express permission. The pilot's ability to see is limited to an arc of about 120 degrees in either direction. Wait for eye contact and a signal such as a thumbs-up. Extreme caution should be used when along the sides of the helicopter, where the pilot may have less opportunity to monitor activity, The tail rotor is extremely dangerous, since it is often less than 6 feet above ground and is invisible when rotating. Unsuspecting people have been decapitated by tail rotors. Approaching the tail end should be completely avoided, except in rear-loading ships, in which case only persons familiar with the equipment should be involved.

Summary

Helicopter safety is basically common sense combined with knowledge about the local equipment. People in positions of authority must strictly control scenes involving helicopter response. The most important elements of helicopter scene safety are knowledgeable personnel, crowd control and choosing an appropriate landing zone. In areas where a specific helicopter is used regularly, the helicopter crew should be asked to provide a familiarization drill with all EMS crews in the area. This could be done at a site that could be used as an LZ during a real incident. In areas where helicopters are used infrequently, it may be helpful to quickly review basic helicopter safety with on-scene personnel if time allows. This is especially important if the scene involves a multiagency response and the person in charge of the scene is unfamiliar with other personnel who are there to assist.

References/Endnotes

1. Dave Samuels, "Smooth Landings: A Safe Approach to Air Medicine," JEMS, November 1989, p.62.

CHAPTER 11
Hostile Situations

Chapter Overview: Some EMS personnel regularly encounter scenes of interpersonal violence. Interpersonal violence is a daily reality for many people. Child abuse, spouse and elder abuse, random and drive-by shootings are all tragic. The purpose of this chapter is to examine the basis of aggression and to introduce ways EMS providers can react appropriately to hostile behavior. Sections on principles of patient restraint and also how to behave if one is taken hostage are included. Basic information about certain weapons, safety at crime scenes, being assigned to stand by at incidents involving crowds or bombs, and other unusual situations, including snipers and extremist organizations is also provided. (Other more common elements of hostility are addressed in Chapters 6,7 and 8.)

Safety hazards associated with aggression are a regular theme for EMS personnel. Hennepin County Ambulance Service, a county hospital-based service in the Minneapolis area found that approximately 20 percent of their 51,729 calls in 1989 involved violence to others, such as shootings, stabbings, assaults and domestic disturbances.[1] Sometimes aggression is still evident when EMS assistance arrives; sometimes it is rekindled while EMS personnel are at the scene. In the Hennepin County study, approximately 10 percent of the staff suffered reportable line-of-duty injuries as a result of the "violent actions of patients, family members or bystanders."[2] Guns, knives and other weapons increase the potential for severe injury Obviously, EMS personnel should request police assistance when needed, especially to addresses known as unsafe from prior experience.

People Who Harm Other People

Many elements of the modern world conspire to bring out the aggressive tendencies in people. Some people are better able to control this emotion than others. Some people set out intending to cause harm, and may have already targeted their victims. Others may become violent because their tempers have exceeded their self-control, sometimes resulting in harm for someone in the wrong place at the wrong time. Because EMS personnel arrive quickly after the event, they may become targets. Aggressive people can-and do-harm others, even arriving caregivers.

Some would argue that people in a medical emergency cannot be expected to act rationally The nature of crisis can change people, usually in one of three ways:

* *Regressively.* Returning to an earlier emotional stage of development, like the tantrum-like activity seen in a toddler.

* *Depressively.* This does not usually involve outright aggression. However, one's state of mind can change. One patient who started with a depressed affect erupted minutes later into a 250-pound mass of fury.

* *Aggressively.* Built-up adrenaline fuels a flurry of violence, often stemming from emotions such as anger and frustration. Another emotion that can alter a scene dangerously is grief. When telling others that a loved one has died, the EMS provider may encounter a violent response. Appropriate interpersonal techniques can help prevent dangerous outbursts.

An interesting phenomenon in emergencies is the tendency for people to "re-freak."[3] That is, after the initial surprise of a medical emergency, many people can cope to some degree while help responds. But when surrendering oneself or a loved one to the care of strangers, some people lose control again. Most dangerous are those whose aggressive response to crisis resurfaces abruptly after prehospital care has begun. With the EMS providers empathic assistance, most people can achieve rational behavior. This makes working with such people more pleasant, and also safer.

People may be aggressive because of medical or self-induced states, such as drug impairment. PCP, crack, "ice," LSD and any other mood-altering street drugs may generate dangerous behaviors. Certain medical problems, such as hypoglycemia, can cause violent behavior. Five strong, youthful emergency responders struggled to restrain one elderly diabetic until a paramedic was able to inject dextrose. the patient was a nice, gentle person when his blood sugar returned to normal. Despite unintended and medically related behavior, though, the patient had previously been potentially harmful to the caregivers.

Suicidal people may be so intent on their mission that they don't care who gets in the way. For example, in 1993 a Denver firefighter was shot and killed by a person who had been reported as suicidal. The police had been called to a possible suicide and had been unable to get a response from the house. They requested that the fire department gain access through a second floor window. The firefighter used a ladder to get to the window. When he reached the window the suspect shot him. Everyone had assumed that the suspect either had already taken his own life or was not in the house. For this reason, it is important never to let a suicidal person out of view. Some may ask to go to the bathroom; have them wait until they get to the hospital. Allowing them out of immediate contact may give them the chance either to. finish the suicide or to obtain a weapon to use against would-be helpers.

Some individuals intending to hurt others may be career criminals with precise information and training. Their prison time may have been spent studying police tactics[4] and learning methods of evasion and confronting authorities. Individuals who remain hostile after release from prison may still pose a risk to authorities, including EMS personnel. Part of the problem lies with the demographics of caregivers: EMS professionals are altruistic and tend not to be physically or emotionally prepared for fighting with the people they are committed to help. Yet the prehospital world is full of tough people. "Resistance is commonplace among today's offenders. About half of the people arrested challenge police authority at least verbally, At least 12 percent are violent or aggressive, requiring 'coercive contact.' Today's average [police] officer, however, is smaller, weaker in upper body strength and more middle-class-rooted than ever before. Lacking the rough and tumble experience of contact sports or military experience, many have never even been punched hard, certainly have never been in a fight for their lives. Yet they're expected to subdue people who . . . were accomplished street fighters by age 6 and who've pumped iron to physical perfection in prison."[5] EMS personnel are exposed to many of the same people as police officers and must take care.

Various cues can help the EMS provider determine the level of threat posed by patients and bystanders. Learning to read the body language of other people quickly and correctly is an essential skill for prehospital care.[6] There are three ways to respond: positively, negatively, or somewhere in between. "Yes" behavior is when someone responds cooperatively both verbally and non-verbally. "Yes" people do what is asked of them willingly and appropriately, "No" behavior may be something like using an expletive and defiantly grabbing the arms of the chair when asked to move over to the stretcher. Both "yes" and "no" behaviors are easier than "maybe" behavior,

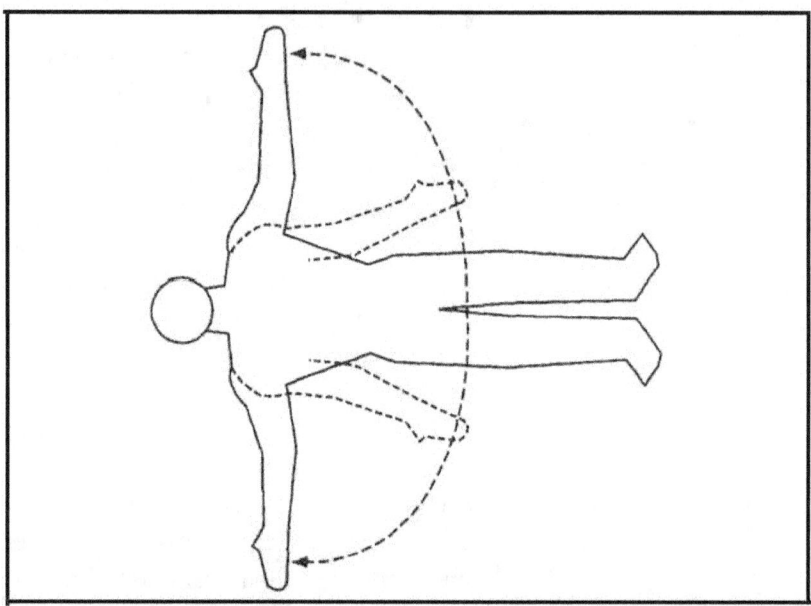

Figure 11-1: Positioning oneself anywhere in the zone shown by the arrows is risky when working with a person lying down.

Figure 11-2: The EMS provider can still see the patient's feet in his peripheral vision while looking the patient in the eye. At this distance, most people do not have legs long enough to kick the EMS provider. If the patient steps into a kick, the alert professional should have enough distance to step back safely.

which is somewhere in between and involves unpredictable verbal and non-verbal messages. "Maybe" people experiment to see how EMS providers will respond to their behavior before deciding whether or not to cooperate. They may be expert at manipulating the situation to their best advantage. "Maybe" people must be watched especially closely.[7]

Stay out of reach until achieving an acceptable level of confidence about personal safety. An EMS provider who gets too close, too fast is at

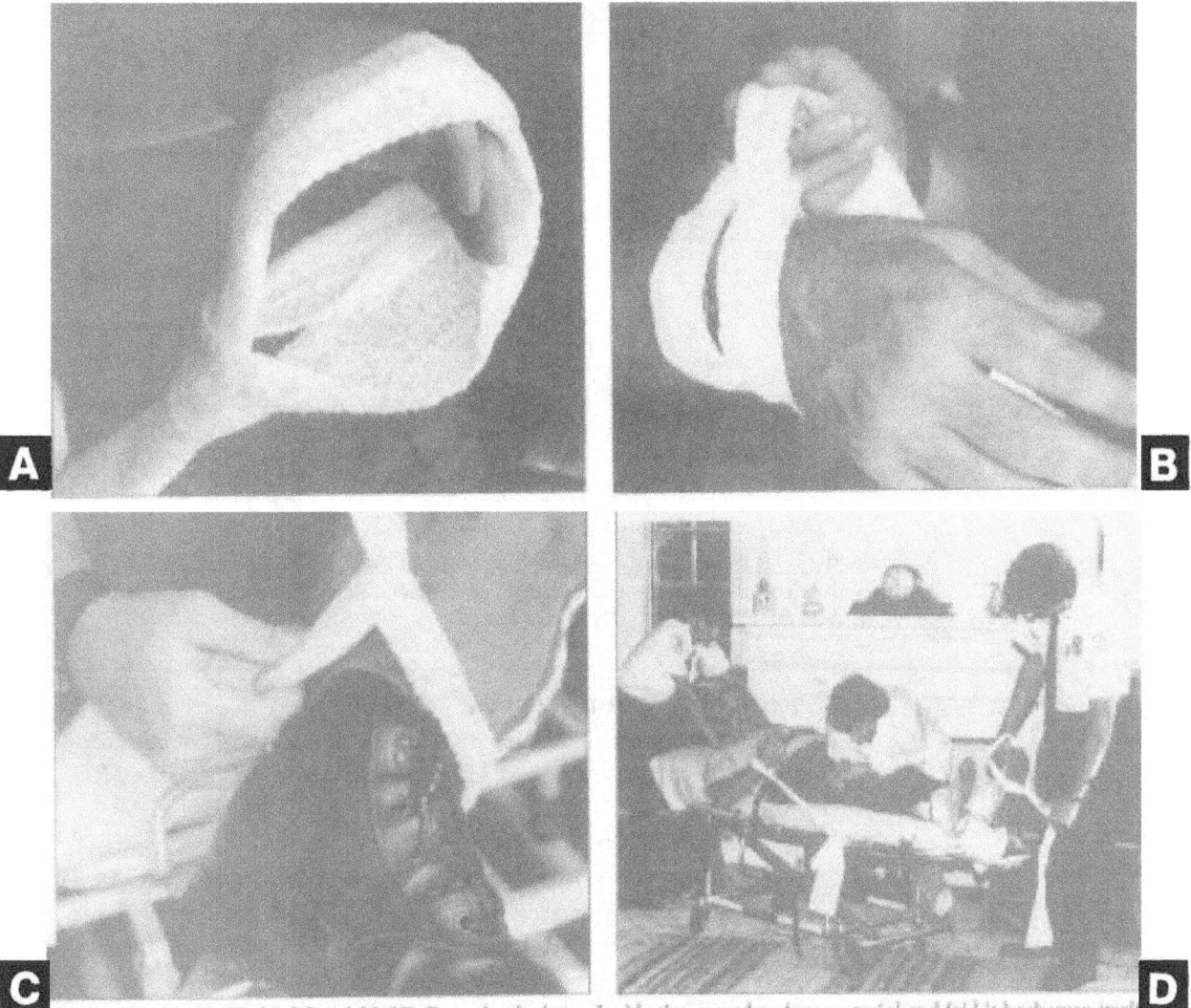

Figures 11-3A, 11--3B, 11-3C and 11-30D: To make the loop, double the gauze bandage material and fold it back upon itself. This loop will tighten on the patient-and can be loosened to provide relief once the patient settles down. Use strong gauze material. Slip the loop over the hand, and tie the ends out of reach. A large loop can be made to slip quickly over the patient's foot regardless how large it is.

risk for being grabbed by panicked or aggressive people. Especially dangerous is being within the half-circle created by outstretched arms on a standing person or in the area of risk shown in Figure 11-1 on a person lying down.

Use peripheral vision to gauge how far back to stand to avoid being kicked. If one can see the other persons feet with peripheral vision while looking into the other persons eyes, one is outside normal kicking range.[8] (see Figure 11-2) If the other person takes a step, the alert EMS provider who stands with one leg ahead of the other should still be far enough away to push off and back away. (NOTE: This technique may require additional distancing for people with extra-long legs.)

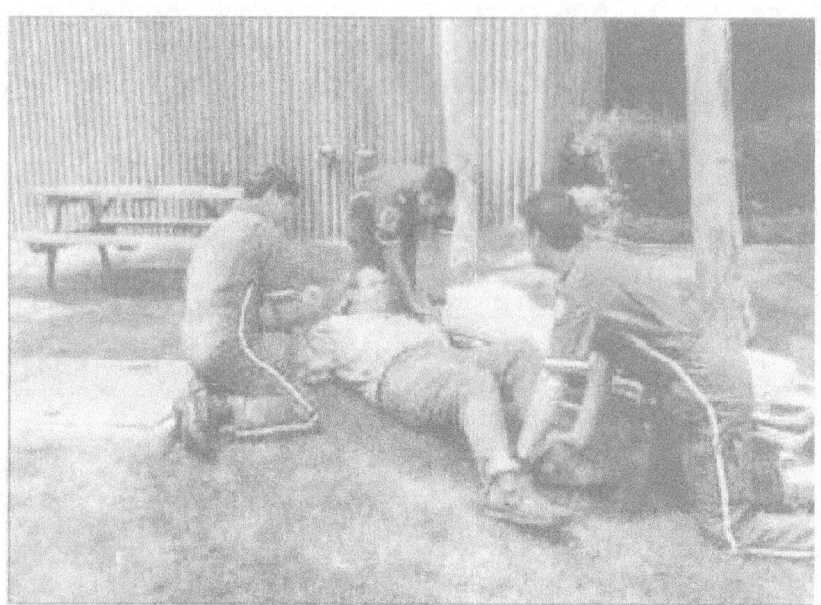

Figure 11-4: Holding the patient at the wrists and ankles, as shown in this simulation, is an invitation for a strong patient to fling "rescuers" about the scene like rag dolls. Hold the patient's knees and elbows instead.

Principles of Patient Restraint

When patients pose a continuing risk for harm to themselves or others despite the EMS provider's best efforts, restraint may be necessary. Restraints are a last resort and must be used judiciously Basic principles for safe patient restraint are:

* *Make sure patient restraint is allowed.* Jurisdictions have their own laws and rules concerning allowable restraint. Local protocols and policies should be written. Know them. If rules are considered inadequate by field providers, work through the appropriate channels to change them properly

* *Make sure patient restraint is justifiable.* Make every effort to avoid using restraints. This might help avoid accusations of brutality and kidnapping. Restraints are a last-resort option, not for inappropriate reasons such as to get away from a scene quickly in order to get off duty on time.

One rookie paramedic was called to a mentally disturbed woman standing in the corner of her empty living room; she had moved all the furniture to the lawn. Efforts to engage her in conversation failed, so the paramedic decided the patient would need to be restrained. Her partner said, "Shall we see if she'll walk with us first?" They approached her from each side, and as she continued her loud litany, she was gently guided by the elbows to the EMS unit. People with bizarre but not aggressive behavior may often be transported safely without restraints. Stay alert for changing behavior.

Some people-especially those who feel suicidal or aggressive-may actually appreciate being restrained. Restraint eliminates the necessity for self-control. Try asking certain patients, "Would you feel better if you were in restraints?" Do not be surprised if the answer is a relieved "Yes!"

* *When possible, defer restraining to law enforcement officers.* Law enforcement officers are trained and equipped to handle the job of restraint.

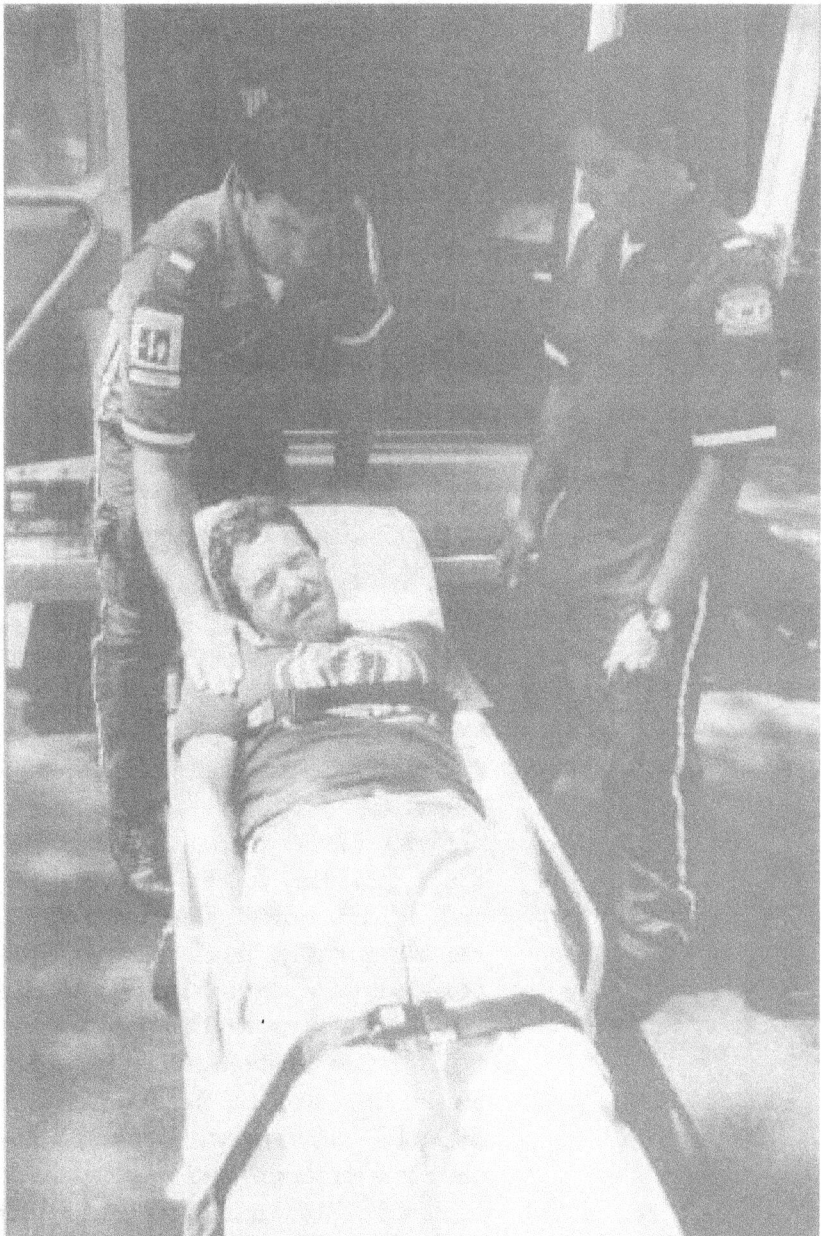

Figure 11-5: In addition to gauze restraints, properly buckle the patient with belts under the arms and over the chest, at the distal upper leg, and (ideally) at the hips. Be sure arm restraints are out of the patient's reach.

Defer to the appropriate ranking officer, and assist in any way possible.

• *Know the equipment.* Methods of restraint vary, and range from strong gauze bandage rolls in half-hitch knots to leather or Velcro restraints. Whatever is used locally, know how to put it quickly and confidently on a struggling patient. Be sure the site of attachment to the stretcher allows for quick release without giving the patient the chance to undo the restraints, (see Figures 11-3A, 11-3B, 11-3C and 11-3D)

• *Have a workable plan.* Coordinate and communicate with those involved. There should be at least one person per extremity. DO NOT grasp the ankles or wrists; instead, grip the knees and elbows. This prevents

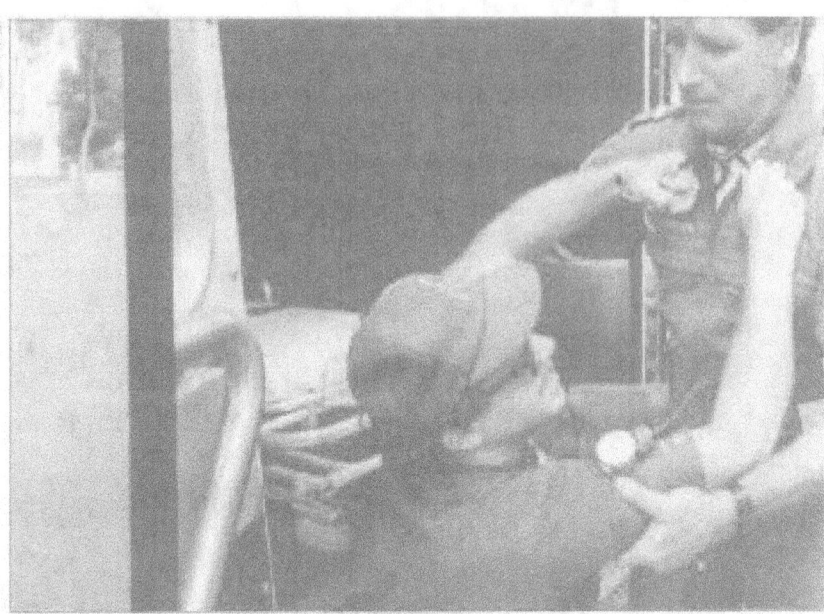

Figure 11-6: The EMS provider's own equipment can be used against him. Be cautious about giving others an easy opportunity to cause harm.

being flung about like the end of a whip. (see Figure 11-4) Be sure the stretcher is in place, with the head turned correctly and with belts ready to be buckled.

Decide whether to place the patient prone or supine. The prone position is particularly effective when someone is totally out of control because it limits movement better and places the patient where the airway can drain naturally. Tie all four extremities and use the seat belts. Some people put the scoop stretcher over the patient to further limit movement. CAUTION: Monitor the ABCs especially carefully.

The supine position improves chances of communication once the patient settles down and makes certain medical intervention such as starting an IV and measuring the blood pressure easier. Run the straps of the chest belts under the arms near the armpits, then over the chest. If the hands are tied down, this configuration prevents the patient from wriggling down and sitting up (see Figure 11-5).

If handcuffs are used, do not place the patient supine. Lying on handcuffed hands can cause circulation to the upper extremities to be cut off, causing injury, Alternative positions are prone or on the patient's side. Check circulation to the hands frequently regardless of which restraint system is used.

• *Be dispassionate.* Restraint must always be done in a non-punitive manner. EMS providers are human, and hostile behavior can be infuriating. Good judgment can be clouded if someone lashes out, spitting, kicking and trying to bite. Such moments require good judgment and a steady professional demeanor. A takedown procedure should never be done revengefully.

• *Never remove restraints in the field.* People who need to be restrained often settle down during transport. However, their behavior is unpredictable, and they may become aggressive again. Ease any discomforts possible-but never untie a restrained person in a moving EMS vehicle.

There are times when a patient is so calm upon arrival at the emergency department that the staff cannot relate to the threat that preceded use of restraints. Certain patients-such as those high on drugs-may be volatile and they may erupt again at any time. Communicate this to the staff so they avoid making false assumptions about the need for restraint.

People With Weapons

A weapon may be anything that can be used for protection, In most cases, though, people focus primarily on knives and guns when thinking about weapons; one New York store alone sells an average of 20 guns daily (and when the 1992 riots in Los Angeles began, stores sold 300 a day).[9] More important than the weapon is the intent of the person holding it. If intent to harm becomes evident, the EMS provider must try to retreat and summon law enforcement assistance.

The EMS provider may discover a concealed weapon during the head-to-toe survey. Depending on the circumstances (such as the patient's level of consciousness and your comfort level handling the weapon), either remove the weapon or try to continue with the exam calmly until the weapon can be properly taken away. To abruptly back off in surprise and horror gives that patient a chance to gain control of the weapon-and perhaps use it.

During the remaining medical exam, be alert for additional weapons. Where there is one weapon, there are often others. Some people have been found to be carrying up to 10 or 15 knives and guns, concealed in every location imaginable, including:

- in the collar at the back of the neck
- in one (or both) axilla (armpit)
- in a brassiere
- anywhere along the waistband
- on the belt buckle
- in holsters hidden under outerwear, especially on the thigh
- in pockets and purses
- strapped to the wrist and hidden by shirt sleeves
- stuck in socks, leg holsters or boot tops.

Medical equipment has also been used against EMS personnel, including small oxygen tanks and equipment from cabinets. EMS providers who keep stethoscopes around their necks should be aware that a hostile patient could decide to use it as a strangulation device (see Figure 11-6).

Ideally, law enforcement officers should handle weapons, making it rare for an EMS provider to need to handle them. If it becomes necessary, carefully pick up the gun by the trigger guard (the curve of metal that protects the trigger) and place it out of the way Never poke anything such as a pen into the barrel of a gun since this could destroy important criminal evidence. Do not let it point at anyone. Document for the police where items of violence were found and where they were moved.

Sometimes, a person in police custody is handed over to EMS personnel for transport to a medical facility. It is easy to assume-often incorrectly-that the patient has been checked for weapons and is unarmed. Unless the EMS provider witnessed the police frisking procedure personally, use an especially careful head-to-toe exam for the added purpose of weapons discovery. The exam may even turn up the occasional dirty hypodermic needle. The physical exam is emphatically not a frisking procedure.

A patient in handcuffs is not necessarily under control. Do not be complacent, even if a police officer rides along. Some prisoners are mostly concerned with gaining their freedom. Even if the patient is restrained in handcuffs or patient restraints, there may be access to a hidden weapon, or a handcuff key may be folded in the band of a leather belt or hidden under a Band-Aid.

Crime Scenes

Efficient EMS providers learn to use assessment of a scene for a variety of reasons. All the senses should be routinely alert to notice factors relevant to 1) safety, 2) medical intervention priorities and 3) interpersonal communication (the 1-2-3s of EMS).[10] Thus several important processes of discovery are happening at the same time.

At a crime scene, these skills can be put to additional use. Crime scenes are potentially dangerous. In many cases, weapons were (or may continue to be) used. Emotions are intense. Substance abuse, particularly alcohol, is common. Career criminals may know harmful evasion tactics. In the case of going to a criminals location-such as a drug production laboratory-there may be homemade security systems, guard dogs or at least people who are armed and feeling threatened.

Try to avoid inadvertent disruption or destruction of criminal evidence. In one case, EMS was called for an unconscious person in a garage. The report was of a patient who had fallen head first off a small ladder, sustaining a head wound. By the time anyone suspected the injury was caused by a bullet, nearly everyone had walked carelessly through the four inches of fresh snow outside. Ample evidence had been destroyed.

EMS providers can help interagency relations by understanding the nature of a crime scene and what can be done to preserve evidence. The important principles are:

• Avoid moving furniture or anything else at a crime scene except to provide vital medical care. Report any changes to the scene verbally to on-scene officers and note it in the written report.

• If relevant, make a crude outline of the patient's body with medical tape. Alternatively, many EMS units now carry instant cameras; a photo can help investigators examine exactly how the patient was positioned.

• Do not cut through bullet holes when scissoring clothing. Ballistics experts may rely on minute evidence available at the site of a bullet hole-but only if it remains intact. Be gentle and careful with clothing, which

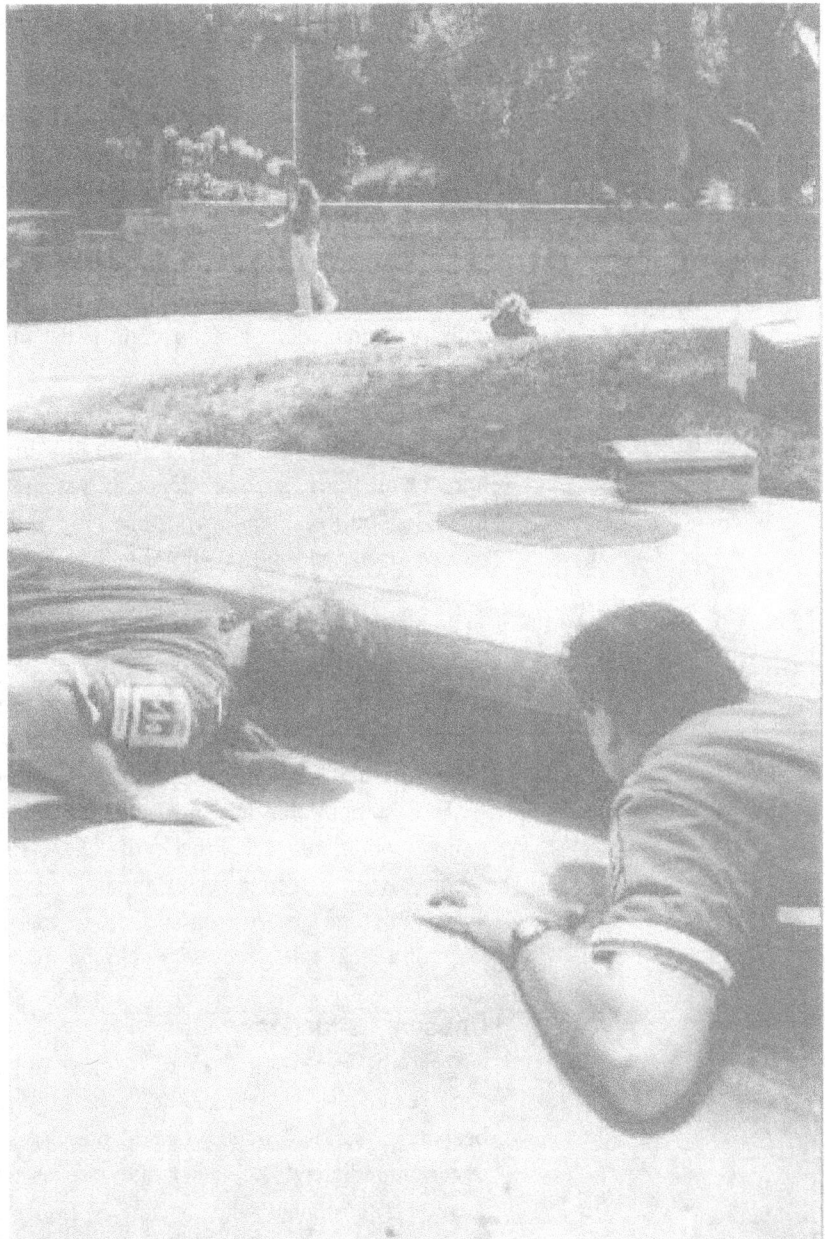

Figure 11-1: If the curb is the only cover available, use it

may contain valuable evidence, especially after a shooting. Do not shake or turn it, and place it in a paper bag.

* Preserve other trace evidence. Bullets, bullet fragments, shotshell wads (large wads of material which are part of a shotgun shell) and other particles of evidence may be enough to convict someone-if they are handled properly. The science of forensic ballistics is now so sophisticated that tiny pieces of evidence can be matched with the weapon from which it was fired. Keep such evidence in a plastic (non-metallic) container, such as an emesis basin or plastic bag.

Chain of evidence is an important concept to law enforcement officers who attempt to bring criminals to justice. Rules of the chain of evidence require that evidence never be outside the direct custody of the person responsible for it. It takes a lot of effort to secure a conviction, including handling evidence properly so it can be used in court. No wonder law enforcement personnel are sometimes so intent on priorities that interfere with those of prehospital providers; when enough evidence is inadmissible, a criminal may go free. Not only does this frustrate the law enforcement officers, but it may also mean the EMS provider may have to deal with a criminal or his victims again in the future.

Standby Posting

EMS personnel are sometimes asked to stand by at law enforcement situations of all types, such as sniper or hostage incidents and other situations involving police. Standby posting may be necessary during civil disturbances, riots and other environments involving large crowds. Standby assignments can drag on for hours-and then break loose in moments. Maintaining an appropriate mental edge is challenging.

EMS providers are usually held at a staging area behind the scenes, Avoid the temptation to sneak up closer to see what is happening. Once summoned, be prepared to respond quickly In certain circumstances, a National Guard or police department escort may be necessary.

At a standby for a bomb site, use special caution regarding use of radios. Sometimes, a bomber will set the first bomb to attract emergency agencies with a second set to explode once they are at the scene. Bombs have sometimes been programmed to detonate when someone keys the mike of a local emergency service radio frequency.

Unusual Situations

Snipers. In some cases, an EMS crew might find themselves in the kill zone before the dangers of the scene become apparent. Moving to safety becomes a survival issue. General principles when on foot include getting cover-immediately! Look for the closest adequate protection. If getting there entails running with the line of fire, move in a zig-zag, erratic fashion to create a more difficult target. Try to maintain a low profile. Sometimes, the best protection consists of a prone position against the curb (see Figure 11-7); for increased protection, try to get next to a drain or sewer grating, where there is usually a dip in the gutter. Even a fire hydrant may work; conform as much as possible to the shape of the cover. Be careful: at one sniper incident, a police officer felt safely covered by concrete pillars; from the snipers point of view, however, the officer was easily noticeable and killed with one shot.[11]

Stay behind cover until the scene is secured. The EMS providers best source of defense at a time like this is the portable radio. Trapped providers have been known to leave secure cover in order to try to reach the radios in the emergency vehicle. A portable radio eliminates the frightening sense of isolation and should be standard issue.

Hostage Incidents. Being taken hostage is another unlikely-but not impossible-hazard of working in EMS. People who tend to take hostages are those who:

• Have a mission to promote. These are similar to the hostage-takers in the Middle East who held a number of Westerners for up to six years in the 1980s and early 1990s.

• Get caught in the act of committing a crime and grab a convenient hostage as a bargaining tool. This could happen to any EMS provider who casually walks into a store for a mid-shift snack and interrupts a crime in progress.

• Have a mental health problem. According to the FBI, 52 percent of hostage incidents involve mentally disturbed people.[12]

One of the most terrifying elements of being a hostage is all the unknowns. What is known is that one's very life is at stake. The average length of a hostage incident is between 4½ and 5 hours, but even short-lived hostage incidents have a powerful psychological influence on most people, even when the outcome involves no physical harm.

If taken hostage, the best tactic is to remain calm. This is easy to say, but hard to do. During the initial ("capture") phase of the incident, hostage-takers are at peak emotional states-excited, nervous, quick to react violently. The calmer the hostage, the sooner the hostage-taker will feel in control and settle down.

Movement or transport of the hostage is also a particularly vulnerable time. During any change of location, do not struggle; follow captor demands quickly and completely. Staying calm also enables one to notice evidence-how many steps were climbed, and whether the location is near a reliable landmark such as train tracks or an airport. This may help later with police investigation,

For action-oriented personalities such as EMS personnel, being a model hostage is a challenge. The most appropriate hostage behavior lies in the middle of a continuum that runs from cowardly to daring.[13] This means behaving patiently during the holding stage while authorities negotiate a safe release. Do not challenge the hostage-takers, or stare at them; do nothing to become a burden or threat. Speak normally. In fact, the more inconspicuous one can be, the better. This may include minimizing the look of authority that can be associated with an EMS uniform. Remove insignia, nameplates, even the uniform shirt, if possible. This will also minimize the tendency among other hostages to view the EMS provider as someone who should take charge of the situation.

Accept concessions or favors offered by the hostage-takers. This may indicate that they are starting to view the hostage as less of an object and more of a person, which is good. Rest whenever possible; sleep is a healing state.

Forceful resolution of the incident is extremely dangerous; in the confusion, it can be difficult to differentiate hostages from hostage-takers. Should law enforcement officials decide to rush the scene, lie flat and still until instructed to move. Throughout the incident, attempt to stay clear of doors and windows.

Extremist Groups. There are many special-interest groups that may pose a danger to people in authority Extremist groups may have political, religious or social views that permit or even encourage very dangerous behaviors. They tend to be steadfast in their beliefs and group practices. Many have arsenals of weaponry, and they are well-trained and unafraid to use them. Others engage in practices that result in member desensitization to horrific interpersonal violence-from gang rape to sacrificing animals and babies.[14]

Sometimes, extremist groups are the object of police action. If they feel cornered, they will often fight back viciously, Their training and beliefs leave them with little or no conscience about who is challenging them and why. It seems obvious that the EMS provider must stay outside the kill zone. However, there are numerous examples of foolhardy behavior in such instances. In one raid against an urban extremist group, the fire department had been called to try to flush out the suspects with fire hoses; they were placed in the line of fire, with only their fire hoses for cover.[15]

EMS personnel may be called to assist law enforcement personnel at scenes involving extremists and their activities. Rural EMS personnel may actually encounter extremist groups relatively more often, because extremist groups often seek the seclusion afforded by a rural setting. There would not usually be a reason to interrupt a group living in such relative isolation. They are free to conduct war games, sadistic rituals and other activities.

EMS personnel are unlikely to be the first to arrive at situations involving extremist groups unless there is an unrelated call for medical aid (or unless they are being purposely set up). For example, a fractured leg on the farm 15 miles out of town may be the result of activities related to training for the group's political mission-but all the EMS provider is allowed to see is the injury itself. Knowledge of the true activities surrounding the incident may not be honestly explained. These groups are jealous of their privacy and will do anything to preserve it.

However, turning a blind eye to the presence of these groups is similar to wishing a local hazardous materials incident will never happen. A safer approach for the EMS provider may be to seek information from local law enforcement personnel and other resources about what groups exist locally.

Summary

Aggression is encountered frequently by people in EMS. There are many interpersonal tactics available to the EMS provider to help calm scenes of interpersonal violence and aggression. These include appropriate word choices, body language, facial expression, proper distancing and presence. A professional's ability to use these tactics may be trust-building and sometimes literally lifesaving.

Law enforcement personnel are prehospital brethren. EMS personnel share many of the same dangers and risks as people in law enforcement. Yet the missions of each group vary, and the priorities of each are often different. Joint appreciation for the training and expertise of each service must replace the misunderstanding and mutual intolerance that can develop. One way the EMS provider can cultivate good relations-and create a safer working environment, both on individual scenes and generally-is to try to protect the crime scene and preserve evidence.

Most EMS personnel will never encounter cases involving snipers or hostages, extremist groups or cults. However, no safety manual would be comprehensive without at least mentioning them and suggesting tactics for coping.

References/Endnotes

1. Robert A. Ball, "EMS Under Siege: A Tale of Four Cities,"JEMS, April, 1991, p.59.

2. Ball, p.59.

3. Jeff J. Clawson and Kate Dernocoeur, Principles of Emergency Medical Dispatch (Englewood Cliffs, NJ: Brady, 1988), pp.72-73.

4. Ronald J. Adams, Thomas M. McTernan, and Charles Remsberg, Street Survival: Tactics for Armed Encounters (Northbrook, 111: Calibre Press, 1980). p.36.

5. Charles Remsberg, The Tactical Edge: Surviving High-risk Patrol (Northbrook, 111: Calibre Press, 1986), pp.423-425.

6. Kate Dernocoeur, Streetsense: Communication, Safety and Control, see Chapters 3 (Understanding Others) and 4 (Effective Communication).

7. The "Yes-No-Maybe" person concept was presented at a "Street Survival Seminar," conducted by Mike Taigman and Bruce Adams and sponsored by Calibre Press, Chicago, Illinois, May, 1988.

8. Taigman and Adams, 'Street Survival Seminar," Chicago, Illinois, 1988.

9. New York Daily News, May 15, 1992.

10. Kate Dernocoeur, "Total Patient Care" day-long workshop, EMS Today conference, Tucson, Arizona, 1990.

11. Adams, Street Survival, pp. 166-67.

12. Fuselier, GW. "A Practical Overview of Hostage Negotiations. FBI Law Enforcement Bulletin 50:2, June/July 1981.

13. Dennis R. Krebs, When Violence Erupts, p.129.

14. Larry Kahaner, Cults That Kill, pp. 16-23 and 234-236.

15. Taigman, Adams, "Street Survival Seminar," 1988.

SECTION III
Health Maintenance

CHAPTER 12

Stress Management

Chapter Overview: This chapter examines stress, its causes, signs and symptoms of distress and the typical impact of stress on the EMS provider's family. It suggests proven methods for minimizing or eliminating the effects of stress, outlines maladaptive stress management tactics and examines a form of stress common to EMS known as critical incident stress.

Stress is part of life; without it, people would not feel compelled to attend to various needs such as warmth, sustenance and rest. When daily demands become physically, emotionally, spiritually or cognitively uncomfortable, distress occurs. Someone who says, "I'm stressed out," is actually in a distressed state.

EMS is a very high-stress job. Knowing about stress, how to recognize its signs and symptoms and what to do about it is a safety-related issue. Too much stress hinders performance and enjoyment of a job, has negative effects on the memory and mental stability, can spark interpersonal conflict, increase accident-proneness, absenteeism and attrition and wipe out good morale. A distressed person may take unusual risks and often fails to maintain physical conditioning. Managing stress is essential to good overall health, which, in turn, helps assure one's safety in EMS.

Stress and Common Stressors

Stress is constant. It is not a disease or affliction. It is the nonspecific physical response of the body to the events and conditions of life. It is an ongoing process of adaptation. Too much stress is a leading cause of or contributor to heart disease, hypertension, ulcers, lowered immunity to disease, arthritis, diabetes, cancer, alcoholism, depression and suicide. The cost of stress is enormous. Conservative sources estimate stress costs $20 billion per year to the American economy; some sources extend that figure to as high as $100 billion.[1]

Events or demands upon us that cause stress reactions are known as stressors. A stressor is anything that triggers the stress response. A stressor may be physical, emotional, environmental, financial, spiritual or social. Different stressors affect individuals differently; those that challenge one person may be inconsequential to another. For example, everyday medical emergencies--which are enormous stressors to those who call for help--are not nearly as stressful for EMS personnel, who are eager to handle them.

Common stressors in EMS include administrative hassles, threat of infectious disease or hazardous materials exposure, shift work, repetitive or routine tasks (such as cleaning the EMS unit or running interfacility transfers), inactivity, weather and temperature extremes, traffic, relatively low pay and the exhausting 24-hour nature of EMS. Although physical stressors-the time of day, the temperature and weather, etc.-are real, mental stressors are more insidious and common. While any event may be stressful, one's cognitive and emotional response to it has far greater impact. For example, EMS units have to be cleaned-it's part of the .job. Someone who cheerfully gets on with the task probably finds it less of a stressor than someone who resists and grumbles. This is also true of other unavoidable annoying situations in EMS. How one interprets the events of daily life has much to do with how stressful that life is.

Small stressors may sometimes accumulate into a large stress-related problem. This is cumulative stress. In other situations, a stressor may be sudden and extensive; these are called rapid onset stressors, in which

Sidebar 1:
Physiological Signs/Symptoms of Stress

Physiological responses to a stressor relate to the innate and powerful instinct to survive. No matter how a person responds emotionally or cognitively to a stressor, physiological response is always the same. This is why positive stressors have the same stress responses as negative ones. Signs and symptoms of physical arousal include:

- pupils dilate
- respiration and heart rate increases
- blood pressure increases
- skeletal muscles become tense and strong
- blood glucose increases (to provide energy)
- the ability of the blood to coagulate increases
- perspiration increases
- blood vessels dilate, except for arteries near the skin surface, which contract, leaving a cool, clammy sensation.

These responses allowed the prehistoric human to respond to physical threats--such as turning to fight a charging bull or to flee an attacking leopard. However, the sources of stress in the modern world are often less clear-cut. Most adults can relate to many of the following signs and symptoms when a sudden stressor occurs: pounding heart, sweatiness, rapid and shallow breathing or shortness of breath, shaking, inability to sleep, gastrointestinal upset (including nausea and vomiting or stomach pain), headaches and muscular aches and pains. Many do not even recognize these as stress-related signs and symptoms.

stress levels peak abruptly, causing signs and symptoms of what is known in EMS as critical incident stress.

The effects of stress are physical, emotional and cognitive. Physically, stress is manifested in a universal fashion regardless of the source. This response, known familiarly as the "fight or flight" syndrome, is associated with complex interactions between the endocrine system and the nervous system. (see Sidebar 1) Thoughts are translated into a physical response that is both mechanical and chemical. This mind-body connection begins with perception of a stressor, and is improved or eliminated through similarly complex processes.[2]

The pace of modern life compounds the problem by not allowing a person to rest after a stress response. The physical effects of stress persist. "During exceptionally stressful times, such as when responding to a critical incident or disaster or when you've suffered some severe personal loss, the hormone thyroxine may be released from the thyroid gland. One release of thyroxine, triggered by a single stressor, requires 2 to 3 days to produce an observable effect in our bodies, but its effects can linger in us for 6 to 8 weeks."[3]

The goal of stress awareness is for each individual to know three things. First, what things cause me to feel distressed? Second, what individualized signs and symptoms of distress occur in me? (see Sidebar 2) Finally, which self-care tactics work for me? People are different. What works for one person may not work for another. What works in one period of life may not work later, either. Answering these requires honest self-evaluation.

Sidebar 2:
Signs and Symptoms of Distress

Clues that one has too much stress lie on a continuum that builds upon itself, from mild to life-threatening. Early management is vital for minimizing the effects of stress. Not to notice or act on early clues is to court the possibility of stress-related disease or general unhappiness.

Remember: signs and symptoms of distress vary from one person to another, and sometimes even within the individual.

There are four nearly universal signs of distress:

• *Persistent/fatigue.* One cannot generate a sense of refreshment, even after many hours of sleep.

• *Negativity.* Negativity worsens in those who are normally negative, or it begins in a normally even-keeled person.

• *Cynicism.* Worsens where it already exists and begins where it did not exist before.

• *Diminished job motivation.* If an activity begins to hurt, it is natural to become less motivated to do it.

Physical signs and symptoms of stress:

• general muscular tension
• headache and neck stiffness
• pain between the shoulder blades or in the low back
• mild but chronic respiratory illness
• nonspecific gastrointestinal upset, with chronic consumption of antacids

• feeling "like a coiled spring"
• clenching the jaw or fists

Emotional and behavioral signs and symptoms of stress:

• feeling overwhelmed
• feeling helpless and/or hopeless
• crying easily and without obvious cause
• sleep irregularities (too much or too little)
• eating irregularities (too much or too little)
• increasingly hair-trigger emotions, especially irritability
• overuse of seemingly comforting substances, such as alcohol, caffeine, nicotine, food and prescribed and illicit drugs
• sense of isolation or withdrawal from the world
• feeling relentlessly pressured

Note: some of the signs and symptoms of stress are related to clinical depression. People can become depressed over their circumstances without making the connection to the stressors.

Severe signs of distress:

• major depression
• feelings of persecution
• paranoia
• increased levels of substance abuse suicidal feelings or gestures

One unavoidable stressor in EMS is shift work. Being awake in the wee hours disrupts the body's natural rhythms. It is probably no surprise that the disasters at Three Mile Island, Bhopal (India) and Chernobyl each occurred after midnight. The U.S. Department of Transportation reports that the risk of having a motor vehicle crash is at least 10 times higher between 4 a.m. and 6 a.m. and that as many as 200,000 traffic crashes per year may be sleep-related?

Regular sleeping patterns are difficult for most EMS personnel to achieve. Volunteers, with their regular lives expanded to fit both training and call responses, live with the equivalent of shift work. The worst shift structures, physiologically, are those that confuse the body's natural rhythms and cycles by rotating too frequently. Someone who has to vary between morning, evening and night shifts may develop shift-related signs and symptoms of stress. Avoiding chronic sleep deprivation is a professional responsibility Loss of compassion, hair-trigger anger, diminished judgment and mental acuity are commonly witnessed in people who are too tired. There is no room in EMS for these.

Stress Management Techniques

There are innumerable methods for coping with stress. Some are healthy and helpful; others are maladaptive and destructive. Collectively, the American public has much to learn about stress management; Americans consume more than 20 tons of aspirin per day, and doctors prescribe muscle relaxers and tranquilizers and sedatives to their American patients more than 90 million times a year.[5] Although there is a place for these medicines, their use does nothing for the stress likely to be causing many of the medical complaints. They only ease the pain temporarily-until it is time for the next "quick fix."

The term stress management refers to those tactics that have been shown to alleviate or eliminate stress reactions. With the willingness to persevere and perhaps alter a few habits, implementing appropriate actions is not difficult. "Stress management is not psychotherapy; it cannot take the place of drug and alcohol treatment; it will not ameliorate incompetent or deficit leadership, nor will it eradicate racism and sexism. It will not substitute for [professional] training, adequate safety precautions, protective clothing or necessary medical care. It will not screen deficient applicants or convince public administrators that [emergency agency] budgets should be enlarged."[6] But it can improve one's quality of life-and may even save it.

There are three basic ways to attack stress:

First, consider which stressors can be changed or eliminated. One paramedic decided to ask partners not to use the time between EMS calls for nonproductive griping and complaining. It helped relieve some of that person's stress.

Second, change attitude regarding the stressor. For example, EMS personnel squander considerable emotional energy fighting unavoidable and basically unchangeable conditions such as skid row alcoholics, abused children, nursing home transfers (often viewed as unglamorous) and cleaning filthy, bloody EMS units. They could choose to adopt a more relaxed, philosophical stance.

Third, minimize the physical response to the stressor by employing various stress management techniques, such as a momentary deep breath to settle an anger response, or stretching hourly to minimize the muscular aches and pains associated with sitting in an EMS unit.

Developing an individualized stress management program. First, acknowledge that activities that work for action-oriented, type A personalities may not appeal to the more relaxed, easy-going type B personality Also, the idea of putting self ahead of others is almost heresy to people with rescuer personalities; many EMS personnel must learn that it is all right, indeed important, to be somewhat selfish. To say no sometimes, or take a few minutes every day for "me time" (time solely for personal pleasure) is relaxing and refreshing; this results in a greater ability to give to others. Some people exercise; others read or just sit on the stoop for a few minutes. Try it.

Figure 12-1: As an athletic pursuit EMS is physically demanding. Regular exercise promotes both mental and physical well-being. It also helps minimize the signs and symptoms of stress.

Physical stress management tactics can help an individual change the physical response to stress. They include:

* Exercise. The multiple benefits of physical exercise make it an ideal method for stress management: improved fitness, reduction of buildup of heart disease risk factors, and the general sense of well-being and self-control that comes from physical mastery of an activity (see Figure 12-1)

* Attention to diet and adequate nutrition. This is always an area that can be improved in the emergency services. For more complete information on both physical fitness and nutrition, see Chapter 17.

* Meditation.

* Yoga, or other activities involving slow, stretching body movement, such as tai chi chuan.

* Deep breathing. This is done with the diaphragm, not the upper chest. Deep breathing is inherent to many martial arts and to yoga and meditation.

* Biofeedback. Look for a certified health care professional to teach this.

* Deep relaxation (progressive muscle relaxation).

Training in some of these methods may be found in local community adult education programs. Watch a class or two to learn whether the activity is personally suitable, then try it. If the first does not work out, try another.

Mental stress management tactics are also crucial. It is possible to develop a positive mental attitude, to develop self-assurance and assertiveness, to quit saying "I should," and begin saying "I would really like to . . . ." Many of the destructive mental processes related to stress buildup are traps people set for themselves; by the same mechanism, each person has the mental capacity to alter a lifetime of such habits. This occurs through self-awareness, repetition and effort. There are many self-help books, audio tapes and seminars available to assist the individual interested in learning.

EMS is known as an "absorbing profession"; it is not an activity one does for a few hours a week and then leaves behind. After a while, it can begin to feel as if everyone is making demands for help. When this happens:

• Step back and take a break, such as a vacation or a few days off, or may be just less overtime.

• Develop a non-EMS circle of friends who are not interested in discussing medicine.

• Rekindle an old hobby, or find a new one.

• Spend time with groups of people who are not contending so frequently with emergencies or other crises more pressing, for example, than deciding where to set up the volleyball net.

• Talk with a good friend, or write in a journal; thoughts can become less frightening once they have been brought into the open.

For dealing with the challenge of sleeping at odd times, begin by accepting the fact that shift work is part of EMS. It is possible to minimize many of the stressors surrounding this fact of life. When sleeping during the day, try to find a cool, quiet, dark place to mimic a nighttime environment. To cover the normal sounds of daily life, get a white noise maker, and turn on the answering machine. Some people also post a "Day Sleeper" sign on their doors to dissuade unwanted visitors.

Another tactic is to establish a specific "anchor" time of day which could be viewed as an equivalent for everyone else's "middle of the night." Anchor times are when one should be able to expect to rest without interruption. For a night worker, this may be 8 a.m to noon. On days off, that person might go to bed "early" at 3 a.m. and get up at noon, but may then alter those hours on work days to sleep from 8 a.m. to 3 p.m. Do not try to revert completely to a daytime lifestyle on days off.[7]

Unwind appropriately after a shift. Do not expect to go right home and go to sleep. Exercise should be done a few hours before sleep to allow body temperature to fall back to normal. Avoid heavy meals just before sleeping.

For the best results, stress management tactics must be practiced long-term. The needs of the individual are relevant and important. Be prepared to give at least three weeks to begin to appreciate and reap the benefits of any change. If the effort is not helping by then, permission to stop should be given, avoiding the thought that "this stuff doesn't work," or "I have failed." Once new self-care habits are established, they can become an integral part of each day. In the classical model of the ripple effect, this benefits the individual, the family, the EMS team and the community at large.

Maladaptive responses to stress. These mask the signs of stress without doing anything to heal or minimize them. Maladaptive responses to stress include:

* taking out aggressions on others
* substance abuse
* delaying stress management through such non helpful activities as watching too much TV
* hurtful humor
* suicide

Families and Emergency Services Stress

The passion of EMS touches everyone in the family, whether one is a child, spouse, parent or significant other of someone in the emergency medical services. Any resource to minimize the inevitable stressors must be nurtured and developed. The stresses faced by every family today can be severe. The challenges are not solely due to EMS, but things like the risks inherent to the work, the amount of time it demands, the dedication, and the fear of loss it can generate can strain the fabric of a family Personal relationships are affected by the stress of emergency medical service, especially when the signs and symptoms are not mediated when they occur.

Being an "EMS family" in the eyes of the outside world often means unrealistically trying to appear invulnerable and thus able to "take the heat" when a family emergency (such as a line of duty injury or death) happens. The family must understand that it is acceptable to react humanly and to seek the support and assistance that non-EMS families would normally expect.

Individuals with families are wise to build in group stress-management time. Areas easy to target are:

* respecting one another's activities and interests
* doing certain stress management techniques (such as exercising) together
* educating the family about the transmission of infectious disease
* planning quantity time as well as quality time
* good interpersonal communication.

A natural outgrowth of increased awareness of the stressors brought by EMS upon families are spouse support groups. Some have cropped up informally. EMS conferences have begun to recognize family assistance as a relevant topic for discussion. A model program known as ASSIST was established by the International Association of Fire Chiefs in the mid-1980s. ASSIST stands for Answers for Spouses by Spouses through Interaction, Support, and Training. It is "a mutual help program that harnesses the collective strength and motivation of spouses and organizes a forum for the exchange of ideas and actions."[8]

Employee Assistance Programs

Anyone affected by personal problems has additional stress in his or her life which may hamper job performance. Yet people in the emergency services are expected to cope effectively and continuously with the emotional, mental and physical stresses of life, both at work and at home. This idealistic viewpoint is repeatedly proven inaccurate. EMS personnel-like everyone else in society-sometimes need additional assistance managing their various affairs. Thus the evolution of Employee Assistance Programs (EAPs). These broad-based programs are designed to assist workers with their problems, including alcoholism, abuse of and addiction to other substances, general stress, family concerns, suicidal feelings, legal and financial assistance and concerns about infectious diseases. Programs available to the worker are often extended as well to their immediate families. In addition to providing help for existing problems, EAPs also offer programs intended to educate and counsel people about physical and mental health risk factors in hopes of preventing health problems. Obviously, this can positively enhance one's overall well-being.

Any time a troubled worker can be properly counseled and rehabilitated, everyone benefits. An EAP is both a wise financial investment for the EMS agency and also an important humanitarian gesture. For example,

> "approximately 18 percent of any workforce is losing 25 percent of its productivity as a result of the costs of impaired performance due to alcoholism, drug addiction, and emotional problems. . . Studies have shown that for every $1.00 invested in an EAP the employer will save $5-16.00. These savings can be seen in a decreased use of medical and insurance benefits, savings in worker's compensation claims, fewer grievances and arbitrations, less absenteeism, less use of management time with troubled employees, less employee turnover, and less personnel replacement costs required of training a new employee. Benefits of the EAP are also expressed through the improved morale of the workforce and the rehabilitation of a valuable trained employee and experienced worker. The U.S. Department of Labor estimates the average annual cost of an EAP to the employer ranges from $12-20.00 per employee."[9]

There are different types of EAPs. Some are internal to the organization; others are external. They may be union-controlled. The need for a member assistance program is addressed in NFPA 1500, Standard on Fire *Service Occupational Safety and Health Program*. Two essential fundamentals required of all EAPs are that:

- complete confidentiality be guaranteed, and
- job security or future promotional opportunities remain assured.

The chance for an EAP to be a positive force to counteract the often excessive stresses of the emergency services is a great addition to the growing arsenal of programs that can offer care to the caregivers.

Critical Incident Stress

People in the emergency services are expected to tolerate a certain regular level of crisis. However, some events are unusually emotionally powerful even by EMS standards. These are known as critical incidents, and they can generate certain stress responses. Fortunately, it is now no longer necessary for emergency personnel to tough out the emotional aftermath of unusually difficult situations. A process called critical incident stress management (CISM), developed in the 1980s is available nationwide to EMS personnel.

Events that can typically result in critical incident stress include:

- Traumatic death or disability of a co-worker, especially in the line of duty Also, suicide of a co-worker.
- Traumatic death or serious injury to children.
- Mass casualty events, particularly where there are no or few survivors.
- Prolonged events, particularly those that end with the death of the victim(s).
- Death or injury to a bystander caused by the EMS provider while in the course of providing emergency care, such as hitting a pedestrian on the way to a call for help.
- Events drawing media attention.
- Symbolic events, both on the level of the public at large (such as the Challenger shuttle explosion) and those that are personally significant.

In the interests of long-term health, responses to critical incidents must be faced and properly defused or debriefed. Certain predictable and normal signs and symptoms are common. CISM helps restore the individuals normal equilibrium; denial and avoidance simply prolong the pain, sometimes to the point that EMS personnel leave EMS. Reactions vary but tend to include: [10]

- *Physical:* fatigue, insomnia/hypersomnia, change of appetite, gastrointestinal problems, headache/backache.
- *Behavioral:* hyper- or under-activity, inability to attach importance to things other than the incident, difficulty with concentration, flashbacks, nightmares, memory disturbance, startle reaction, isolating behavior.
- *Psychological:* fear, guilt, emotional numbness, over-sensitivity, anxiety, depression, feelings of helplessness, amnesia for the event, anger (which may manifest by scapegoating, irritability, frustration-especially with bureaucracy) and violent fantasies.

Some of the signs that occur may have to do with the metabolism-raising effects of thyroxine. These are most noticeable when they peak, about 10 to 14 days after a rapid-onset stressor. Without stress education, people are less likely to connect the symptoms with the event. Many think they are just going crazy and fail to seek proper mental health attention. Symptoms of thyroxine release include:[11]

- fine muscle tremors
- worry and/or anxiety
- paranoia
- insomnia
- racing thoughts
- increased internal body temperature

- increased gastrointestinal motility
- increased secretion of digestive juices
- decreased heart strength
- increased probability of heart failure

There are several types of CISM, both at the scene and afterwards. The first, peer defusing, is when a group informally talks over events they experienced together. It tends to occur naturally. Formal critical incident stress debriefing (CISD) is a gathering only of those people who were at the scene. This carefully developed process is designed to assist people in understanding their responses to a critical incident and how to heal well from it. It is inappropriate to try to conduct formal debriefings without proper training; doing so risks causing serious harm to vulnerable rescuers.

A properly formed CISM team consists of both peer supporters (from the emergency services) and mental health workers (who have training in group facilitation and can provide outside referrals to those who occasionally need it). Good teams are regional, have received appropriate training and education, and are non-partisan. A CISM team should never be formed by an organization with the intention of debriefing its own people; total confidentiality and relative anonymity are vital elements of a debriefing.

A third type of CISM is on-site defusion. This is for large, ongoing scenes. People coming from the front lines are rapidly evaluated by a member of the local CISM team for signs and symptoms of distress while they rest before re-entering.

Fourth is demobilization. At a large-scale situation, people are helped by having a buffer period of 30 to 40 minutes before leaving the location. During this time, health-oriented food is available, along with brief education of signs and symptoms of critical incident stress to watch for. The main goal of the demobilization site is to provide a place for personnel to collect themselves before leaving.

National standards have also been implemented for a related process known as Emergency Incident Rehabilitation. The purpose is "to ensure that the physical and mental condition of members operating at the scene of an emergency or a training exercise does not deteriorate to a point that affects the safety of each member or that jeopardizes the safety and integrity of the operation."[12]

Emergency rehabilitation also minimizes, eliminates, or prevents many of the stressors of a large-scale incident. Easily digested fluids and food, such as stew or soup and fruit, are provided. Rescuers must rest for at least 10 minutes and are evaluated medically, (For more in-depth information on how emergency incident rehabilitation can reduce the physiological effects of heat stress and cold exposure, see Chapter 7.)

Many EMS veterans ask why they have never needed CISM before. Didn't critical incidents happen before 1983? Of course they did-at the expense of peace of mind and longevity for many people in EMS. Several 17-year veterans have stories to tell that easily show that no one is invulnerable.[13] Some are haunted years after a critical incident by flashbacks and nightmares. Some quit EMS because of the pain. These reactions are no longer necessary CISM can help anyone in EMS, regardless of years of service.

Summary

Knowledge about the causes and management of stress is especially important for people in emergency services. Some in emergency services deny the effects of both cumulative and rapid-onset stress and believe they are invulnerable. The wise EMS provider acts positively on his/her own behalf. The benefits of a stress management program include:

• Improved overall long-term health, and a greater chance of avoiding stress-related illnesses

• Diminished likelihood of developing maladaptive responses to stress, such as alcoholism or explosive anger

• Greater enjoyment of the work setting, resulting in diminished attrition and improved morale

• Happier family relations.

Remember that "no matter how colorful the bandwagon or how earnest the pitch, stress is not something that can be cured; however, it can be managed as a part of a normal life."14 As Carlos Casteneda said, "The difference between the warrior and the ordinary person is that the warrior views everything as a challenge, while the ordinary person views everything as a blessing or a curse." Stress is here to stay, The question is: Which EMS personnel will tackle it as a warrior would?

References/Endnotes

1. United States Fire Administration, Stress Management: Model Program for Maintaining Firefighter Well-Being, (FA-100, February 1991) p. 16.

2. USFA, pp.26-27.

3. USFA, pp.27-28.

4. Russ McCallion and Jim Fazackerley, "Burning The EMS Candle: EMS Shifts and Worker Fatigue,"JEM.S, October 1991.

5. USFA, p.15.

6. USFA, p.3.

7. McCallion and Fazackerley, "Burning the EMS Candle,"JEMS, October 1991, p. 43.

8. Joanne Fish Hildebrand, "ASSIST Workshops: Mutual Aid for Spouses," Firehouse, September 1985, pp.161-64.

9. International Association of Fire Fighters, Guide to Developing Fire Service Labor/ Employee Assistance Programs, (Washington, D.C.: IAFF, 1992), p.3

10. From "Critical Incident Stress Reactions, by Nancy Rich, MA (Mayflower CISD Team, Denver, Colorado), 1985.

11. USFA, p.28.

12. US Fire Administration, Emergency Incident Rehabilitation. (Washington, DC: FEMA FA-114, 1992.

13. Dana Jarvis, "Around The Nation,"JEMS, February 1985, p.35.

14. USFA, p.43.

CHAPTER 13

Basic Physical Fitness

Chapter Overview: This chapter poses physical fitness as a long-term safety concern necessary to prevent injury or lifestyle-influenced chronic disease. Since EMS can require athletic ability, this chapter explains cardiovascular and muscular fitness along with basic nutrition. Other concerns include smoking, regular physical checkups, hearing and vision protection and issues related to pregnancy in EMS.

Some safety issues take years to manifest their effects. To consider safety to be a concern only of the present moment may prove a narrow view. Physical fitness relates to a longer, healthier life without avoidable chronic disease. On a daily basis, a physically fit body can defend one against on-the-job illness and injury. Not only is the task of EMS, with its many physical demands, easier; it is also more enjoyable.

Physical fitness is the body's counterpart to the mental fitness derived through good stress management. The dimensions of physical fitness that are core to wellness are:

- cardiovascular fitness
- muscular fitness (muscular strength, power, endurance and flexibility)
- body composition
- adequate, healthful nutrition

These components are equally important. If any is lacking, a threshold exists for injury or illness. Someone may have strong muscles but be inflexible and obese. To be fit and unhealthy is possible, but to be unfit and healthy (for long) is less likely.

Prehospital emergency care demands good physical fitness. Compare EMS personnel to a football team. Each loves what they do. Each goes out on their individual "playing fields" and gets continually physically challenged. The difference: all football players train physically for the job, while the majority of EMS personnel readily admit they do not. EMS personnel must learn to view themselves as athletes.

Many EMS personnel also admit to having poor diets composed mainly of substances that contribute to heart disease, stroke, cancer and liver disease. Some pick themselves up with sugar, caffeine and illicit substances. Some use alcohol and cigarettes. There is a powerful sense of invulnerability and denial about the chance of developing a chronic illness. Those who heed the warnings can minimize or prevent many well-recognized and related health problems through long-term physical fitness.

While some EMS organizations specifically encourage physical fitness, anecdotal evidence is that most do not. Personnel are expected to seek adequate and proper exercise programs on their own, so it tends not to happen. Standards for medical and physical requirements for EMS personnel do not exist at this time. NFPA 1582, *Standard on Medical Requirements for Firefighters,* contains valuable information, including guidelines for physicians, which can be used by EMS personnel to establish local criteria. A similar recommended practice on physical fitness/physical performance in the fire service is under development.

In EMS, it has been noted that the muscle groups commonly used are different from those used in firefighting. Frequent lifting from ground and near-ground levels places the emphasis for muscular fitness on different parts of the body than occurs for firefighters. Obviously, people with cross-over responsibilities must train for all the muscular demands of the emergency services.

Cardiovascular fitness. This is the well-known component of aerobic capacity. In a pursuit that demands frequent bursts of cardiovascular effort, it is important to have the ability to bolt into rapid action on a moment's notice. Although one should avoid running on a scene, it may be the only available defense in a dangerous situation.

Aerobic capacity is achieved through regular stimulation of the cardio-vascular system. Good aerobic exercise activities include walking, jogging or running, basketball, aerobic dance, racquetball, swimming, soccer, bicycling and numerous other forms of exercise. Which activity a person chooses is a personal decision and should be based on availability of the necessary equipment and individual interest. The principles involved are:

• raise the pulse to the target heart rate (which is 70 to 80 percent of one's age-dependent maximum),

• for 20 to 30 minutes,

• at least three times per week.

A full medical exam is advisable for anyone preparing to start a fitness program. If the exam includes a medically supervised treadmill test with a continuously monitored electrocardiogram, an individualized target heart rate can be accurately determined. Otherwise, use the following formula:

• Estimate maximum heart rate by subtracting age from 220 (using a 38-year-old person, for example: 220 - 38 = 182).

• Subtract resting heart rate from the estimated maximum heart rate (in this case, resting heart rate is 52, thus 182 - 52 = 130).

• Multiply the figure thus calculated by 0.7 (130 x 0.7 = 91).

• Add the figure just calculated to the resting heart rate (91 + 52 = 143). This is the estimated target heart rate.' When doing aerobic exercise, benefits will be derived at this rate.

Muscular fitness. Within this area of health are the four components of muscular strength, power, endurance and flexibility. Many people in emergency services may be physically fit in some of these areas but not all. Here are the differences:

Muscular strength is the ability to exert force, meaning how much weight can be lifted or pushed. The need for muscular strength in EMS is obvious; one provider had, in a single morning, three patients whose combined total weight was 1,110 pounds!

Muscular power is the explosive application of muscular force, defined as the quantity of work produced in a given time. Power is attributed to sprinters and perhaps also to personnel in the emergency services who are awakened abruptly by loud tone signals and who are expected to be driving to someone else's crisis within seconds.

Muscular endurance is the ability to sustain a muscular contraction for an extended length of time. This is needed by marathon runners and also by rescuers asked to hold a trauma victim's head from an awkward position during a lengthy extrication.[2]

Muscular flexibility refers to the ability to achieve the widest range of motion physiologically possible. Many people who are not flexible mistakenly believe they cannot change. With appropriate daily stretching, increased flexibility can be a major factor in physical fitness. Much low-back pain is due to overtight hamstring and low-back musculature. Never bounce when stretching; this can cause rebound tightening because of the pain that improper stretching can cause. Instead, extend the stretch gently, lengthening muscle fibers gradually.

Total body fitness does not occur just because one is doing plenty of aerobic exercise. Other elements of muscular fitness demand their own specific conditioning. The goal is to put the major muscle groups through an exercise regimen which answers the "Five Rs" (range of motion, resistance, repetitions, rest and recovery) at least two or three times per week. There are many different activities that answer the needs of various muscle groups.

One good all-around form of exercise is properly conducted weight training. Using the entire range of motion on each repetition helps train the muscle to its fullest extent in both directions. Resistance means the amount of weight being lifted. When the correct weight is selected, one can achieve full range of motion, the involved muscles are appropriately tired after the prescribed number of repetitions, progression to higher weights occurs quickly enough to avoid boredom and the lifts can be done without cheating through use of accessory muscles or incorrect leverage.

The number of repetitions to select depends on the individual's goals. Fewer repetitions with higher weights result in increased muscular strength and size; more repetitions with lower weights increase muscular endurance. Each repetition should be done smoothly with the specifically targeted muscles doing all the work. Three sets of repetitions per targeted muscle group are good for both strength and endurance. For the greatest gain, muscles should be pushed to the point of muscular failure. Always have a spotter available when working with heavy weights.

Breathe out during exertion; never grunt or hold the breath to "gut out" a lift. Rest between sets long enough for the muscle to be able to work at near-capacity again; this may require up to two minutes between sets.

Fitness is never achieved overnight. It requires steadfast and consistent effort and a positive, patient mental attitude. Those starting fitness programs often try to do too much too fast. Recovery time is important. A too gung-ho approach to a fitness program may easily sabotage it. Muscles need adequate recovery time between training sessions. The same muscle groups should be exercised no more than three times a week, and never two days in a row. The goal is a training program which can be followed regularly and continually over time.

A common fear among those who love prehospital emergency medical care is back injury Fitness that emphasizes strength, flexibility, power and endurance, is good prevention for low-back injury, Although the true cause of low-back pain is often unknown, onset has been linked to the risk factors of poor lifestyle habits, improper body mechanics, decreased physical fitness levels, poor flexibility and decreased abdominal tone.

In prehospital care, there are certain things one can do to minimize the risk of injury First, don't expect cold, stiff muscles to routinely go to peak performance without adequate preparation. At situations requiring a heavy lift, stretch for a moment; time permits this at all but life-threatening situations. EMS personnel are increasingly taking the time to stretch their quadriceps, hamstrings, and lower backs with 30-45 seconds of stretching before a particularly heavy lift-and they say it helps.

Some EMS personnel are now wearing flex-belts, or more accurately, intertrochanteric belts, which encircle the pelvic ring snugly and provide support in the sacroiliac region. This is especially helpful during heavy lifting. The belt creates improved sitting posture, which is important to people confined to an EMS unit for much of their time on duty. It also minimizes the negative effects of a paunch. It does not create muscular weakness (although constant use without training properly for the task of lifting can promote loss of muscle tone). An intertrochanteric belt is not the same thing as a weight-lifting belt. Proper fitting and instruction in its use are important; some people even advocate that such a belt should only be prescribed by knowledgeable physicians.[3]

Body composition. More than 30 percent of the American public (68 million people) exceed ideal body weight. Of these, 40 million are obese, meaning that they are 20 to 25 percent over ideal body weight. Massive obesity (where one is 50 percent or more above ideal body weight) occurs in three percent of males and nine percent of females. Those whose weight is 20 percent or more above desirable body weight based on height and age are risking their health.

Obesity is directly related to an imbalance of energy expenditure and caloric intake-a state all but a very few can control; only a small minority of obese people can truly blame endocrine dysfunction or genetics. The health risks directly associated with obesity are coronary artery disease, diabetes, hypertension and low back pain. Each is treatable (even reversible, to some extent) with appropriate effort.

One study looked at the issue of age and fitness, initially concluding a negative relationship between the two. However, a second view of the data concluded that it was body fat, not age, that resulted in a loss of performance approaching 50 percent in the 30-year age spread of the subjects. "The impact of body fat on human performance, where body weight is a factor in performance, cannot be overstated. Maintenance of appropriate body composition can enhance performance in the over-40 age group by as much as 30 percent."[4]

The problem of obesity is widespread. The more often and extensively a fat cell is expanded, the easier it is for that cell to re-expand-and fat cells can balloon to 1,000 times their normal size.[5] The problem with constant dieting is that unless one changes one's basic eating habits, pounds lost are usually quickly regained. (see Sidebar 1).

Basic nutrition

Most of us grew up on Mom's constant advice about food: "Eat everything on your plate. " "Drink your milk! " "Have another helping . . . ." Three heaping meals a day in the land of plenty ingrained what are now recognized as poor eating habits-habits that contribute to cancer, heart disease and stroke. In the past 75 years, fat intake has increased 31 percent, and

Sidebar 1:
Rating Diets

With the abundance of diets of the market today, it's important to evaluate a diet before using it. Diet books have been written by many people with various degrees but no formal nutrition training from accredited schools. Ask the following questions of any diet, diet organization or diet book before starting it:

* Does the diet meet nutritional needs for all four food groups without eliminating one entire group?
* Are the required foods inexpensive, easy to find and easy to prepare?
* Can the whole family eat what the dieter eats?
* Does this diet establish good permanent eating habits?
* Is the diet safe, easy to stick to and realistic?
* Does the diet advocate slow weight loss of not more than one to two pounds per week?
* Does the diet refrain from using special pills or formulas to aid in weight loss?
* Is medical advice advocated before starting the diet?
* Is there a maintenance plan?
* Does the diet hold "forbidden" foods to a minimum?
* If joining an organization is required, is the cost of the diet reasonable?
* Does the diet emphasize good food habits and exercise as long-term goals for permanent weight loss?

[Source: USFA: Physical Fitness Coordinator's Manual for Fire Departments, p. 8-19]

sugar intake is up 40 percent. That means that three out of five calories are regularly derived from fat or refined sugar. Although there is nothing innately wrong with a bit of either, the amount eaten can crowd out other essential nutrients and decrease fiber intake.[6] To change one's diet is potentially lifesaving.

Whether the food a person prepares and eats is healthy and appropriate has never been more of a mystery than now, Fewer and fewer people know how to select and cook fresh foods. Instead, they rely on pre-packaged and prepared foods. The concept of altering a diet bound by tradition and habit may seem daunting, but it is not as difficult as it may seem. With some education about the effects of various foods, many people have discovered that a dietary switch is compelling.

Most people grew up hearing about the four basic food groups (dairy, meat, vegetable, grain), and believed that eating plenty of each of those wholesome foods was healthy. Re-examination has led nutritionists to refute the habits of a generation. Too much dairy food and meat is unnecessary and probably harmful. Emphasis is now on fiber-rich and water-rich food instead of cholesterol- and fat-filled foods. In fact, the Physicians Committee for Responsible Medicine has developed a new way to categorize the four basic food groups: grains, legumes, vegetables and fruits[7] None of these foods has any cholesterol; their percentage of calories measured as fat (with a few exceptions, such as nuts, avocado, olives and oatmeal) is less than 11 percent. Most meat and dairy products are heavy in both cholesterol and percent of calories measured as fat; they are considered food choice options by the same physician committee.

It is also important to hydrate appropriately. One should consume six to eight glasses of liquid daily. Those laced heavily with sugar, caffeine, additives or alcohol are less nutritious or potentially injurious. Regular soda pop contains 12 to 15 teaspoons of sugar, and diet soda pop contains additives (the long-term effects of which are still unsubstantiated). In the late 1980s soda pop replaced water as the number one liquid consumed in America; annual average per capita intake is 490 cans or bottles. Plain water is the most wholesome fluid to drink, and it is easy to obtain and cheaper, too.

Some types of cancer- the number two cause of death in the U.S.-are diet-related. "At present, we have overwhelming evidence . , . [that] none of the risk factors for cancer is...more significant than diet and nutrition."[8] Chief culprits are meat and fat. All populations worldwide with a high meat intake have high rates of colon cancer. Foods identified as cancer causers include broiled and barbecued meats, fried foods, hot dogs, ham and pickles, beer, wine and hard drinks. Those identified as cancer "cancelers," or foods that should minimize the risk of cancer, include anything high in vitamins A and C (found in salads and vegetables), cruciferous vegetables (particularly broccoli, cauliflower and brussels sprouts), and fresh, crunchy green, red and yellow vegetables. The National Cancer Institute suggests at least three servings of a half-cup of vegetables (or one cup of leafy vegetables) and two pieces of fruit per day.[9]

Be careful of fast food. It is full of sugars, fats and salt. Sodium intake alone is a source of concern, given that one in three Americans has hypertension. Nutritionists would be happy if Americans dropped sodium intake to 1,100 to 3,300 milligrams (½ to 1½ teaspoons); the body only really needs 200 mg (1/10 teaspoon). Yet the average American consumes 4,000 to 6,000 mgs per day.[10] One source advocates eating at fast food restaurants no more than once a week, because the number of calories tend to be out of proportion to the nutrients.[11]

Good eating habits are challenging when one's schedule is as erratic as that of the typical EMS provider. Instead of stopping at a fast-food place, seek out the neighborhood grocery store and buy some fruit and juices. Plan ahead, and bring a whole-grain sandwich and fruit or cut vegetables in a cooler; progressive EMS agencies are beginning to provide coolers on the EMS units for this purpose. Instead of buying so many soda pops, stop in at the hospital and get a large cup of ice water-it will last longer and refresh better.

As nutritional science becomes increasingly sophisticated, the evidence is difficult to ignore. What a person eats has a compelling correlation with long-term health. The confusion that reigns leaves many people frustrated, resulting in a "head in the sand" attitude toward food. Assistance can be gained by reading or from a toll-free nutrition hotline where registered dieticians can answer questions: 1-800-843-8114.

Additional concerns

Smoking: The surgeon general identified smoking as a health hazard in 1964. Since then, smoking has been implicated in a number of chronic illnesses, including respiratory illness. Non-smokers who spend time in secondhand smoke (the smoke caused by other smokers) are in a hazardous atmosphere as well. Smoking is considered a primary risk factor for coronary artery disease, directly related to how many years and how heavily one has smoked. It causes elevated blood pressure. It reduces oxygen-carrying capacity of the blood with its by-product of carbon monoxide, thus placing an extra burden on the circulatory system. And it is believed to contribute to plaque formation in the coronary arteries.[12] Smoking cessation is an obvious and well-known requirement for good overall health.

Regular Physical Exams: Many people employed in the emergency services are required to have physical exams for licensing or recertification. Many people regard this policy as a nuisance rather than as the potentially important procedure it is: to provide for preventative medicine and to document work-related illnesses or injuries and to seek out subtle medical findings.

A regular physical exam should be done by a qualified occupational physician who understands the demands of EMS and the pride with which its professionals serve. An occupational physician should know options and treatment opportunities to return workers to the workplace.

Hearing protection: A normal conversation creates about 65 decibels of sound (dBA). EMS personnel are frequently exposed to much higher and more sustained noise levels, sometimes exceeding safety limits recommended by the Occupational Safety and Health Administration (OSHA). Many of these noise levels cannot be controlled, which means that hearing protection, not noise abatement, is the preventative measure of choice. Those who protect themselves against noise-induced hearing loss are more likely to enter old age with better hearing than those who do not.[13]

One EMS-related study sought to calculate the total time of siren exposure for 184 adult males, who had been screened for occupational and recreational exposure to noise (from firearms or heavy machinery), past or current use of toxic drugs, medical history, and family history of hearing loss. Their length of employment averaged 6.436 years, with ages ranging from 21 to 59 years. The study concluded that "duration of siren noise exposure in this busy EMS system correlated significantly with degree of hearing loss," despite non-job related noise exposures. "We have shown that the hearing loss is accelerated at 1½ times the rate expected for age-matched controls."[14]

Hearing loss relates both to levels of noise and duration of exposure. Siren speakers installed on the front grille of EMS units (rather than on the roof) reduced one common source of noise for EMS personnel.

Ear Plug and Ear Muff Comparison

Feature	Earplugs	Earmuffs
Noise Protection	Varies	Greater, less variable
cost	Less expensive	More expensive
Size	Small, convenient	Not easily carried or stored
Fit	May be difficult	One size fits most
Monitoring Use	Can't be seen	Easily seen
In Hot Environment	Comfortable	Uncomfortable
Head Movement	No restriction	Restrictive in close quarters
Wearability	Only in healthy ear	Can be worn even with minor ear infection. Can interfere with glasses, head gear and hair.

Adapted from United States Fire Administration, Fire & Emergency Service Hearing Conservation Program Manunal (FA-118), p. 31; Federal Emergency Management Agency, Emmitsburg, MD

Figure 13-1: Hearing conservation is an important longtern wellness issue. Be careful to protect hearing in some way when exposed to loud noise.

Regardless of where siren speakers are placed, the ASTM standard on hearing protection mandates that noise levels in the front cab shall not exceed 85 dBA.[15] Similarly, the National Fire Protection Association recommends a maximum limit of 85 dBA without audible warning devices and 90 dBA with warning devices in operation. However, there is significant noise exposure elsewhere for field personnel, such as generator-powered extrication equipment, street and highway noise and screaming people. As a generally youthful group, EMS personnel seldom focus on hearing loss as a major personal concern. However, wearing hearing protection at any time the environment is noisy does matter to long-term wellness. The EMS provider can minimize hearing loss by wearing hearing protection. It does not interfere with the ability to hear radio transmissions. There are several different types: military-type foam inserts, ear plugs and flight-type ear muffs. The NFPA recommends protective ear muffs, due to the difficulties of proper fit and insertion of ear plugs.[16] (see Figure 13-1) One 30-plus-year EMS veteran kept his ear plugs in his front pocket so that they were always handy. Others hang them on a string around their necks.

Eye protection: Vision is also too precious to jeopardize. Despite encouragement to use eye protection at scenes with blowing dust or flying debris, many EMS personnel do not wear protective gear for the eyes. The effects of ignoring this self-care opportunity could be long-term. Regular, proper eye examinations are important.

Reproductive Issues in EMS: As increasing numbers of women enter prehospital care, the issue of pregnancy and safety becomes more apparent. A number of factors related to pregnancy pose issues of safety for the mother, her partner and the people being served. The physiology of pregnancy, reduced agility and fleetness, and loosening pelvic and low-back ligaments are of concern. Add to that the customary vulnerability EMS professionals face due to the unpredictable and sometimes violent prehospital environment. Whether a woman should continue to work while pregnant is an individual choice protected by law; some women can and do work into the third trimester.

Women are protected under the Pregnancy Discrimination Act of 1978 (PDA), which is an amendment to Title VII of the Civil Rights Act of 1964. That act states, "Women affected by pregnancy, childbirth or related medical conditions shall be treated the same for all employment related purposes, including the receipt of benefit programs, as other persons not so affected but similar in their ability or inability to work . . ." (This applies to employers with 15 or more employees and to employment agencies and labor organizations.)[17]

In one national survey, 74 percent of EMS organizations had no pregnancy policy.[18] Those that do need to adhere to certain findings from a U.S. Supreme Court case that was decided in March, 1991. In the UAW v. *Johnson Controls case,* the issue was whether employers have the right to bar women from hazardous occupations in order to protect an unborn child (and themselves from potential lawsuits) should the woman become pregnant. The Supreme Court ruled that employers may not exclude women from jobs in which exposure to toxics might harm a developing fetus.[19] That and any other medical concerns for relieving a woman from duty must be related to concerns for the woman's health, not that of the fetus.

Whether light duty should be mandated and even whether a woman should be required to report a pregnancy remain unanswered questions. The *Johnson Controls* case has helped clarify that it is not legal to force pregnant EMS personnel into layoff status if no light duty assignments are available. The pregnant EMS provider must be treated as well as any other member of the personnel roster who is injured off-duty.[20]

Summary

Physical fitness is a lifelong project. In this age of instant gratification and immediate solutions, the idea that fitness is hard to acquire and easily lost can be overwhelming. There are many elements of fitness-strength, flexibility, nutrition, smoking (and other bad habits) cessation. However, those who make the effort to achieve fitness are more likely to live healthier and happier lives. In the end, old age will conquer even the healthiest person. Yet there is one encouraging concept: "While we cannot control our chronological age, there is good evidence to suggest that our physiological age may be significantly reduced through positive lifestyle strategies, including exercise, managing stress, not smoking, and modifying dietary intake."[21] Those who heed the warning and act upon the evidence in their youth are most likely to reach old age in shape to enjoy it.

References/Endnotes

1. USFA, Physical Fitness Coordinator's Manual for Fire Departments (FA-95), p. 9-13

2. USFA, p.9-21,

3. David Imrie and Lu Barbuto, The Back Power Program (Toronto: Stoddart Publishing Company, Ltd., 1988). pp.99-101.

4. USFA,p. 1-7.

5. Kate Dernocoeur, "Big Bodies,"JEMS, September 1986, p.47.

6. Michael Jacobson and Sarah Fritschner, The Fast-Food Guide (NYC: Workman Publishing, 1986).

7. News item, MS Magazine, May/June 1991, p.83.

8. B. Reddy. "Nutrition and its Relationship to Cancer, Advances in Cancer Research 32:237, 1980.

9. Peter Jaret, "Eating Your Crucifers Really Can Cut Your Chances of Getting Cancer," In Health, Sept/October 1991.

10. Jacobson, Fritschner, Fast Food Guide, p.64.

11. Jacobson, Fritschner, Fast Food Guide, p.28.

12. USFA, p.5-2 and 5-3.

13. NFPA, NFPA 1582: Standard on Medical Requirements for Fire Fighters, 1992, p.20.

14. Paul E Pepe, et al, "Accelerated Hearing Loss in Urban Emergency Medical Services Firefighters," Annals of Emergency Medicine 145, May 1985, pp.438-442.

15. ASTM F1230-89, "Standard Specification for Minimum Performance Requirements for Emergency Medical Services Ground Vehicles,", 1989.

16. NFPA, NFPA 1500: Standard on Fire Department Occupational Safety and Health Program, 1992, p.38.

17. USFA, FA-128: A Handbook on Women in Firefighting: The Changing Face of the Fire Service, January 1993, pp.43-44.

18. Kate Dernocoeur and James N. Eastman, Jr., "Have We Really Come A Long Way?: Women in EMS Survey Results,"JEMS, February 1992, pp. 18-19.

19. EMS Insider, "Supreme Court Ruling on Fetal Protection Affects EMS Maternity Policies," May/June, 1991.

20. Ibid.

21. USFA, Physical Fitness Coordinator's Manual for Fire Departments, p. 5-5.

CHAPTER 14

Infection Control

Chapter Overview: This chapter discusses infectious diseases and how they can be transmitted, prevention opportunities (on scene, at the station and through proper immunization), basic hygiene and how to properly clean medical equipment.

The safety issues outlined in this manual cover a broad range of topics. Some of the safety measures encouraged among prehospital providers will prevent sudden injury or death. The protective effect of others is more long term or subtle. The latter applies to the protective effect of infection control.

General infection control awareness is a relative newcomer in the prehospital consciousness. The AIDS epidemic has had a powerful collective impact; its lethality is a major reason for significant changes in patient care habits that have occurred since the early 1980s. However, it is also important to be aware of the other infectious disease risks, especially hepatitis B. In fact, there is a much higher risk of being infected with HBV (hepatitis B virus) than HIV (human immunodeficiency virus, which causes AIDS).

Because of the intensity of interest in infection control, many educational programs, reading materials and mandatory education programs have been developed. This overview is intended to serve solely as a primer. Some information changes in the discipline of infection control; cross-check facts and use only currently reliable, accurate and in-depth sources of information.

Infection transmission and common infectious diseases

Medical personnel are exposed to a broad scope of diseases. Some illnesses are a nuisance; others are serious and even lethal. Fortunately, exposure is not the solitary mechanism needed to acquire an illness. The following five factors must intersect in order for an infection to occur:

1. *Dose,* or the number of live organisms present.

2. *Virulence,* the strength of the infecting organism (the ability to infect).

3. *Host resistance,* or the ability of the host to resist the effects of the infectious organisms.

4. *Route of exposure,* such as airborne, bloodborne, or foodborne.

5. *Means of transmission,* or a way for the organism to gain entry into the host.

Infection can be prevented by interrupting the disease process at any of these points.[1] This can be accomplished by following good infection control practices and by understanding diseases and their transmission.

Two of the most serious communicable illnesses that EMS personnel must protect against are AIDS and hepatitis B. To highlight the difference between dosages, consider the following: one cubic centimeter of human blood may contain one to 100 human immunodeficiency viruses (HIV), the organisms that cause AIDS. The same volume of blood may contain one hundred thousand to one billion hepatitis B viruses.[2]

There is also a dramatic difference in virulence. HIV can only be 'transmitted from direct blood-to-blood (or certain other body fluid) contact. However, the hepatitis B virus can remain infectious for weeks on smooth, dry surfaces such as the aluminum stretcher bars. "Although the idea of getting hepatitis B may not hold the terror that AIDS does, it is nonetheless a very unpleasant disease that increases the potential for liver cancer and cirrhosis of the liver later in life."[3] Hepatitis B, too, can be fatal.

Other communicable diseases include hepatitis (types A, C, and other non-A non-B), meningitis, tuberculosis (which is staging a strong comeback), salmonella, herpes, malaria, polio, rabies, even lice and scabies (surface bugs that cause skin irritations), and impetigo (a staph or strep infection of the skin). Childhood diseases such as chicken pox, mumps, measles and whooping cough are also increasingly prevalent. (see Sidebar 1)

Infectious diseases are transmitted in various ways, including:

• *Via contact.* There are several types of contact. Direct contact is direct physical contact between the infected person and a susceptible person. For bloodborne diseases, direct contact must occur with blood or body fluids. Or sometimes the susceptible person has to come into direct contact with the pus, discharge, or other secretions directly related to the infection. Indirect contact occurs when the infected person touches an object, and a susceptible host then touches the contaminated object. Droplet contact occurs when a person coughs or sneezes; airborne droplets containing infectious organisms may be sprayed from the mouth or nose. Droplets may remain airborne for a long time, and may spread over fairly large distance. Illnesses spread this way include tuberculosis and influenza.

• *Via an intermediate host or vector:* Some infectious diseases are spread via ticks, mosquitoes, flies, or fleas. Lyme disease, Rocky Mountain Spotted Fever, malaria, and other illnesses are spread this way,

• *Via other vehicles.* Infectious disease can spread through water, food and other vehicles. For example, hepatitis A is spread through food.

One way to prevent infectious disease is through immunization, Fortunately, vaccinations and boosters exist for many infectious diseases. Certain infectious diseases once thought eradicated are once again on the rise; one reason is the complacency that can occur when people do not witness the devastation of these illnesses because they were inoculated as children. Other reasons include that some people lack access to primary health care, and many immigrants never had the chance for immunization.

EMS professionals should take advantage of all immunization and booster opportunities. It is required by law in certain cases. Even for illnesses that are not life-threatening, immunization can eliminate much of the anxiety of being around diseases such as the infectious childhood diseases. Some vaccines require boosters, including tetanus and diphtheria. Understand the relevance of immunization in terms of avoiding infecting patients; in particular, rubella (German measles) can be devastating to a pregnant woman's fetus. In addition to rubella, tetanus, and diphtheria, vaccines are available for influenza, polio, measles and mumps. The vaccine for flu is established yearly after the likeliest strains are predicted; over time, an annual flu shot will provide immunization

Sidebar 1: Disease Information for Emergency Response Personnel			
Disease/Infection	**Mode of Transmission**	**Vaccine Available?**	**Signs and Symptoms**
AIDS/HIV (human immunodeficiency virus)	Needlestick, blood splash into mucous membranrs (e.g., eyes, mouth), or blood contact with open wound	No	Fever, night sweats, weight loss, cough
Chickenpox	Respiratory secretions and contact with moist vesicles	No	Fever, rash, cutaneous vesicles (blisters)
Diarrhea: Campylobactor Cryptosporidium Giardia Salmonella Shigella Viral Yersinia	Fecal/Oral	No	Loose, watery stools
German Measles (Rubella)	Respiratory droplets and contact with respiratory secretions	Yes	Fever, rash.
Hepatitis A (Infectious Hepatitis)	Fecal/Oral	No	Fever, loss of appetite, jaundice, fatigue
Hepatitis B (HBV) (Serum Hepatitis)	Needlestick, blood splash into mucous membranes (e.g., eye or mouth), or blood contact with open wound. Possible exposure during mouth-to-mouth resuscitation	Yes	Fever, fatigue, loss of appetite, nausea, headache, jaundice
Hepatitis C	Same as Hepatitis B	No	Same as Hepatitis B
Hepatitis D	Same as Hepatitis B, dependent on HBV (past or present) to cause infection	No	A complication of HBV infection and can increase the severity of HBV infection
Other non-A, non-B Hepatitis	Several viruses with different modes of transmission (These are called "non-A," "non-B" because there are no specific tests to identify them.)	No	Fever, headache, fatigue, jaundice
Herpes Simplex (Cold Sores)	Contact of mucous membrane with moist lesions. Fingers are at particular risk for becoming infected	No	Skin lesions located around the mouth area
Herpes Zoster (Shingles) localized disseminated (See Chickenpox)	Contact with moist lesions	No	Skin lesions

Disease/Infection	Mode of Transmission	Vaccine Available?	Signs and Symptoms
influenza	Airborne	Yes	Fever, fatigue, loss of appetite, nausea, headache
Lice: Head, Body, Pubic	Close head-to-head contact Both body and pubic lice require intimate contact (usually sexual) or sharing of intimate clothing	No	Severe itching and scratching, often with secondary infection. Scalp and hairy portions of body may be affected. Eggs of head lice (nits) attach to hairs as small, round, gray lumps
Measles	Respiratory droplets and contact with nasal or throat secretions Highly communicable	Yes	Fever, rash, bronchitis
Meningitis: Meningococcal	Contact with respiratory secretions	No	Fever, severe headache, stiff neck, sore throat
Haemophilus influenza (usually seen in very young children)	Contact with respiratory secretions	No	Same as above
Viral Meningitis	Fecal/Oral	No	Same as above
Mononucelosis	Contact with respiratory secretions or saliva, such as with mouth-to-mouth resuscitation	No	Fever, sore throat, fatigue
Mumps (infectious parotitis)	Respiratory droplets and contact with saliva	Yes	Fever, swelling of salivary glands (parotid)
Salmonellosis	Foodborne	No	Sudden onset of fever, abdominal pain, diarrhea, nausea and frequent vomiting
Scabies	Close body contact	No	itching, tiny linear burrows or "tracks," vesticles-particularly around finger, wrists, elbows and skin folds
Syphilis	Primarily sexual contact; rarely through blood transfusion	No	Genital and cutaneous lesions, nerve degeneration (late)
Tuberculosis, pulmonary	Airborne	No	Fever, night sweats, weight loss, cough
Whooping cough (pettussis)	Airborne, direct contact with oral secretions	Yes	Violent cough at night, whooping sound when cough subsides

Adapted with permission USFA, Guide to Developing and Managing an Emergency Service Infection Control Program (FA-112), (Emmitsburg, MD: FEMA, 1992). pp.14-18

for a wide variety of viruses.

Because of the seriousness of hepatitis B, the Occupational Safety and Health Administration (OSHA) has mandated that vaccination be made available to all employees in clinical contact with patients within 10 working days of assignment.[4] This includes all covered EMS personnel, as well as students on rotation in medical settings. OSHA estimates that universal vaccination of at-risk employees would prevent from 244,000 to 274,000 cases of HBV infection over 45 years, resulting in the saving of some 5,400 to 6,100 lives over that time.[5] Health care workers have the right to refuse a hepatitis B vaccination, but a specially worded document must be signed. There is no vaccine for HIV

Should personnel be exposed, especially to HIV, OSHA also mandates that EMS organizations have a process of treatment, referral and counselling available. Careful attention must be given to preserving patient confidentiality; at the same time, EMS personnel should not be denied appropriate care for illnesses contracted in the course of duty.

On-scene Operations

OSHA has clearly written criteria for exposure control.[6] The standard written by OSHA sets requirements for limiting exposure to those workers at risk through a combination of engineering controls, personal protective equipment, and worker training.[7] It can, and does, heavily fine organizations that do not adhere appropriately to the rules surrounding this important safety issue. NFPA 1581 is the infection control standard for fire departments; in many ways, it exceeds the OSHA requirements by incorporating both exposure control and infection control. The protection offered by the OSHA document extends to all "at risk" employees, defined as anyone who may be reasonably anticipated to come into contact with human blood or other potentially infectious body fluids. All emergency response personnel meet this criteria.

There is no way to predict in the short time frame and general confusion common to EMS operations which infectious diseases might be present. Therefore, EMS personnel should use body substance isolation (BSI) against potential infection. BSI goes beyond universal precautions and applies to all body fluids. Wearing gloves, mask, gown, and eye protection, as appropriate, is intended to prevent eye, mouth, other mucous membrane, nonintact skin and parenteral contact with blood, other body fluids, or other potentially infectious material.[8] While blood is the single most important source of HIV and HBV transmission in the workplace, in the field it is safest to assume that all body fluids are infectious.[9] Gloves should be donned prior to patient contact; this is particularly important for EMS personnel with chapped, cut and otherwise non-intact skin on the hands,

Each component of body substance isolation may not always be indicated. The EMS provider should use judgment according to the nature of the incident. If the patient has a persistent cough, for example, the best available protective measure is to mask the patient. The EMS per-

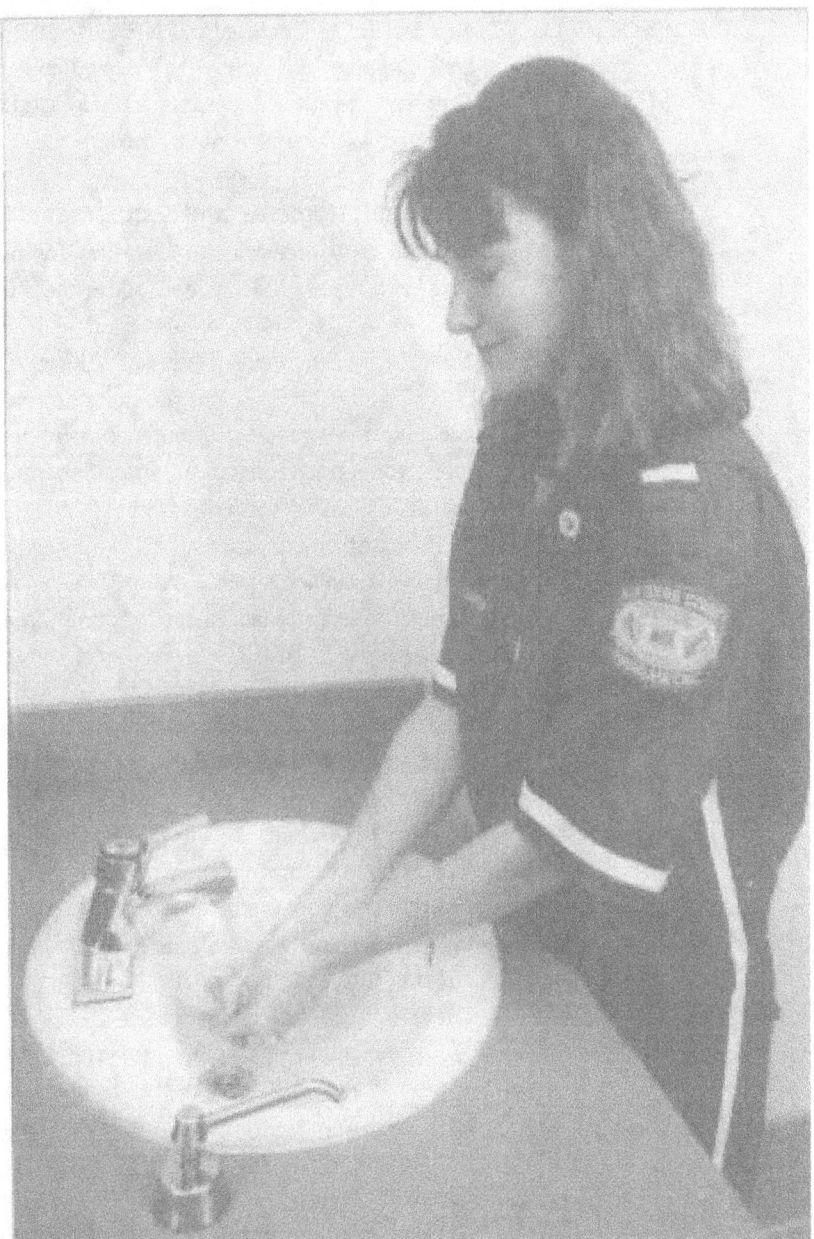

Figure 14-1: Hand washing is known to be the most effective overall infection control measure. Wash properly and often!

sonnel may also need to wear masks or respirators, and ventilate the EMS unit with rapid-moving fresh air. If there is any potential for blood, vomitus or other body fluids to splash, a mask and protective eyewear should be worn. A fluid-resistant gown, apron or coverall should also be considered. It is better to use more protection than less. (Certain uniform materials are now available that allow normal sweat vapor to escape but do not absorb fluids such as blood.)

Needles used in medical care must be properly disposed of into sharps containers that cannot spill or be pierced by the needles within. The jump kits usually taken to the patient's side should also contain appropriate sharps control. Needles should never be recapped or broken. Suppliers have now made available better needles and IV devices that have protective designs to help avoid needle stick injuries; use of these tools should be encouraged.

Personal Hygiene and Equipment Cleanup

In addition to adhering to the rules for body substance isolation, health care workers should always follow the rules of basic hygiene. The best (and easiest) forms of exposure control go back to behaviors learned as children: Wash hands frequently. Keep fingernails clean. Keep fingers out of facial orifices. In addition, behaviors that promote good health-adequate rest, good nutrition, and good general physical fitness also promote good host resistance. These measures minimize host susceptibility should exposure occur.

Handwashing is particularly important. Yet it is frequently ignored, perhaps because it is so basic. One researcher found that medical personnel do appropriate hand washing procedures only about 30 percent of the time.[10] Instill a good habit: wash hands whenever gloves are removed, after all patient contacts, and after disinfecting equipment. (see Figure 14-1) Also wash before eating and after using the toilet. Plain soap is usually adequate for handwashing. The hands should be well-lathered with regular hand soap and scrubbed, using friction, for at least 15 seconds before rinsing well and drying. Because the faucet is notoriously germ-laden, use the drying towel to turn it off.

Proper handwashing facilities are often unavailable, such as is the case at the emergency scene or in situations of non-transport. For such circumstances, antiseptic hand cleaners or towelettes must be supplied on each EMS unit-and used.

Equipment must also be properly cleaned. Cleaning is the process of removing visible surface debris. Disinfecting is the process of inactivating most disease-causing organisms; in most cases, an appropriate mix of bleach and water will suffice. Sterilization is the destruction of all microorganisms in or around an object. This must be done using specialized equipment or chemicals not typically available in the prehospital setting. In EMS, cleaning -usually through old-fashioned scrubbing with warm water and soap-and disinfecting are all that is needed for most equipment. Some equipment (such as endotracheal tubes and OB kits) should not be reused unless it is properly sterilized. Cleanup should be done in an appropriate facility.[11]

Cleaning the EMS unit and other equipment should follow a common-sense approach. There is little need (or time) for lengthy, complicated measures. Wash and disinfect actual work surfaces. This includes the often-ignored stethoscope; in one study, 8 percent of stethoscopes harbored notable infectious organisms.[12] The walls and floors of the EMS vehicle should be washed regularly, Remove surface debris with sweeping, then scrub with warm, soapy water. Disinfect with an appropriate agent. Good, inexpensive disinfectant and soap products are available; germicidal and viricidal agents should be approved by the Environmental Protection Agency (EPA).

Recommendations regarding washer-dryer facilities for linens and clothing, as well as facilities for appropriate clean-up and disinfection, are generally provided in such documents as *USFA's Guide to Developing and Managing an Emergency Service Infection Control Program and NFPA 1581, Standard on Fire Department Infection Control Program*. Contaminated disposable equipment and personal protective equipment, as well as biohazardous waste, are now referred to by OSHA as "regulated waste." This waste should be legally disposed of according to local laws and regulations.

Summary

Minimizing the risk of infection is one of the few controllable aspects of prehospital emergency medical care. Although no effort is failsafe, awareness of the problem and preventative strategies will serve one well. The mission of EMS is to care for people who are sick or injured, but there is no reason emergency care providers should become sick in the process.

References/Endnotes

1. USFA/National Fire Academy, Infection Control for Emergency Response Personnel: The Supervisors Role (Student Manual), (Emmitsburg, MD: FEMA, 1992) pp. 3-8.
2. Martin Favero (Centers for Disease Control), in panel discussion, NAEMT Conference, Reno, Nevada, June 1988.
3. Kate Dernocoeur, Streetsense: Communication, Safety and Control 2nd ed. (Englewood Cliffs, NJ: Brady, 1990), p.269.
4. Occupational Exposure to Bloodborne Pathogens: Final Rule, 29 CFR Part 1910.1030 (Washington D.C.: Federal Register, December 6, 1991), pp. 64175-64176.
5. Richard M. Duffy; Clifford S. Mitchell; Sharon Doyle; James Melius, Infectious Diseases in the Fire and Emergency Services, (Washington, D.C., International Association of Fire Fighters, 1992), p.19.
6. OSHA 29 CFR part 1910.1030, p. 64179.
7. Richard M. Duffy; Clifford S. Mitchell; Sharon Doyle; James Melius, Infectious Diseases in the Fire and Emergency Services, (Washington, D.C., International Association of Fire Fighters, 1992), p.19.
8. USFA, Guide to Developing and Managing an Emergency Service infection Control Program (FA-112), (Emmitsburg, MD: FEMA, 1992). p.108
9. USFA, Guide to Developing and Managing an Emergency Service Infection Control Program (FA-112), (Emmitsburg, MD: FEMA, 1992). p.68
10 Katherine H. West, infectious Disease Handbook for Emergency Care Personnel, (Philadelphia: J.B. Lippincott Company, 1987).
11 NFPA 1581: Fire Department Infection Control Program, 1991 Edition, sections 3-1 and 3-2.
12. West, Infectious Disease Handbook, p.79.

SECTION IV
Manager Responsibilities

CHAPTER 15

Safety and the Manager

Chapter Overview: The manager's role in safety must be advocated. This chapter addresses loss control and risk management with regard to the cost savings and reduction of employer liability. Adherence to state and federal safety rules depends on knowledge of them. Written protocols and procedures must be established and enforced, and appropriate interagency relations developed and maintained.

Among the numerous issues faced by EMS managers in the 1990s is the broad topic of safety Although exact statistics are lacking, it is probable that EMS receives more than its fair share of the 10,000 work-related deaths and 6 million work-related injuries annually. It is not enough to say "stay safe out there" and think the responsibility for safety has been shifted successfully to those on the streets.

Although it may not seem a compelling management priority, safety awareness and follow-through by smart managers can have important impact. Knowing that personnel are avoiding serious (even fatal) injuries is obviously important. However, another important driving force is the impact of certain actions on the organization's financial bottom line and to employer liability For example, one back injury (the most common work site injury reported) averages $6,000 in direct and indirect costs.[1] Appropriate attention to safety will increasingly prove to be both a life- and cost-saving measure.

Pro-active implementation of safety measures will *prevent* certain costs. These savings are often difficult to measure, since they are hidden. For example, if an organization invests company-wide in reversible protective coats, it would be unknown how many people avoided harm because they were either visible or invisible according to the needs of each call. Part of the problem is that when a safety program is working, nothing happens; no one is hurt. Only before and after statistics can help quantify the impact of a broad-based safety program. This is exactly why the manager is pivotal to the success of a safety program; the manager has the authority to create a comprehensive, pro-active safety program.

This manual has described numerous aspects of safety, each of which warrants management awareness. In some cases, federal and state regulations even mandate certain education and protective actions, such as with hazardous materials exposure and infectious disease. Otherwise, EMS personnel have traditionally been left on their own to discover safe prehospital practices through experience. Safety must be the joint responsibility of both field providers and their leaders.

Managers have a large scope of obligation. To ignore management's significant and growing responsibility in the area of safety is to risk professional or organizational disaster. Managers must be willing to demonstrate genuine commitment toward safety and exemplify that commitment through an appropriate attitude.

Not only must a rule be established; it must be enforced. What will happen in an organization where there is a rule mandating seat belt use, yet a provider fails to obey? For one thing, worker's compensation can be lost or reduced; damages awarded to others injured in crashes found to be the fault of the EMS driver may be enlarged; and only partial awards for damages may be awarded the EMS organization even when a crash is found to be the fault of the other driver.[2]

In California-often regarded as a forerunner of national trends-a very comprehensive worker safety law has been passed. Enforced by California OSHA, it is referred to cynically as the "Be A Manager, Go To

Jail" legislation. It places the burden on administrators to provide safety training and to keep extensive records about what is taught and who was there. Penalties are not light and can include felony convictions for managers who fail to abide by the legislation. It is likely to get continually more difficult for managers to rise above accusations of ineptness or ignorance.

When safety standards exist and are followed well, managers can reward that performance. This helps ensure that such performance will be repeated. Otherwise, the only feedback occurs from a negative point of view-that is, when someone gets hurt. The climate of an organization (which is largely within the control of the manager) has much to do with employee job satisfaction. When employees feel that those in supervisory or management positions are genuinely supportive of appropriate safety practices, the inclination to sue for employer liability may be reduced; it will certainly be increased when employees sense disinterest or inappropriate interference by supervisors and management.

The best defense is to know the rules of the game and to implement appropriate educational programs, written policies, and enforcement procedures. Safety management is a prime opportunity to avoid trouble through use of pro-active principles. That way, loss control-of expensive equipment, experienced personnel and even lives-is easier to achieve and acknowledge.

Risk Management and Loss Control

Managers have their fair share of stress and pressure without having more heaped on them by avoidable safety-related problems. Safety is an attitude. It should begin with the highest level of management and percolate through the organization to all areas of the EMS organization. What bolsters an honest attitude of safety is identification of risk management opportunities and implementation of appropriate loss control measures.

Risk management is an identification process. It involves recognizing situations which could cause injury or risk to the organization. For example, emergency vehicle crashes are a prime area of concern and are easily identified in the risk-management phase of safety program development. So is pre-employment assessment of applicants, thorough and honest orientation and providing appropriate safety equipment.

One parallel risk-management activity is developing recommendations for shifting risk from a negative (uncovered) risk to one that has been addressed and planned for (a covered risk). This occurs, for example, through appropriate training and retention programs, sufficient insurance coverage and a host of other areas deserving of attention. Another risk management activity (and one that can help highlight the cost savings enjoyed by organizations which implement these programs) is maintaining statistics and reports on types of injury, complaints, lawsuits and insurance claims.

The best loss control programs are preventative in nature; they involve creating programs designed to reduce anything-such as vehicle crashes and other incidents-that will affect the health and well-being of employees and their equipment. For example, implementation of a driver training and performance standard using low-force driving, along with medical priority dispatching to reduce use of lights and sirens are loss control activities. Loss control also establishes procedures for what happens *after* a risk has become a reality, so that the effects of the problem can be properly managed and minimized.

Proof that attention to risk management and loss control affects the bottom line comes from one interstate trucking company This company was experiencing a string of crashes-with associated rising insurance costs. It took the threat of bankruptcy to raise safety to be a major priority. Once this happened, they were able to reduce their losses tenfold. "Daily attention to all areas and in particular potential problem areas was the key in this case. Business always improves along with the safety record; injuries decrease, costs go down, and productivity goes up." At the bottom line, this trucking company, through management-supported implementation of safety programs, decreased insurance costs by $13.4 million, incidents by 33 percent and the cost of worker's compensation by $7.2 million.[3] Similar benefits are in store for EMS organizations that choose to make safety a top priority

Writing Policy and Procedures

Policies and procedures are the.employee's guidebook to a job. Although it may be possible for employees at startup organizations to "write their own rules," future generations of employees will need a guidebook of rules to understand the way things are done.

Approaches to establishing policies and protocols take many forms. In one major Asian city, visitors were informed that field personnel would never have to wonder what to do in a given situation, since every conceivable prehospital situation had been addressed in the protocols. The manager then turned proudly to the wall of his office and showed the visitors books many inches thick and weighing many pounds. At the opposite end of the spectrum is another company's values-driven protocol; written by a California EMS organization, it is the size of a double-sized business card folded in two and contains less than 300 words. Most companies take an approach to writing protocols that lies between these extremes. (see Figure 15-1)

The purpose of rule-making is to establish baseline guidelines for appropriate behavior and medical care and to explain the disciplinary actions that are taken when those guidelines are violated. Many of the actions taken daily by prehospital personnel are done according to their integrity, Integrity involves having awareness of what is right and wrong-and doing right. However, written protocols are necessary in an age when lawsuits are quick to be filed if something goes wrong. If an EMS crew leaves a scene, for example, because they perceive a significant threat to their safety (the right thing to do in an organization

---

STAR CARE Checklist

The following is a checklist you can use to analyze almost any patient care issue you might encounter. Go through the list in order from top to bottom, and ask yourself if your care meets each criterion. If it does, chances are that you can defend your actions in almost any forum.

Customer-accountable

If I were face-to-face right now with the **customers** 1 dealt with on this response, could I look them in the eye and say "1 did my very best for you."

Appropriate

Was my care **appropriate-medically,** professionally, legally and practically, considering the circumstances 1 faced?

Reasonable

Did my actions **make sense?** Would a reasonable colleague of my experience have acted similarly, under the same circumstance?

Ethical

Were my actionsfair **and honest** in every way? Are my answers to these questions?

Safe

Were my actions **safe**- for me, for my colleagues, for other professionals and for the public?

Team-based

Were my actions taken with due regard for the opinions and feelings of my **co-workers,** including those from other agencies?

Attentive to human needs

Did 1 treat my patient as a **person?** Did 1 keep him/her warm? Was 1 gentle? Did 1 use his/her name throughout the call? Did I tell him/her what to expect in advance? Did I treat his/her family and/or relatives with similar respect?

Respectful

Did 1 act toward my patient, my colleagues, my first-responders, the hospital staff and the public with the kind of **respect** that 1 would have wanted to receive myself?

Reprinted with permission from Baystar Medical Services/Thom Dick

Figure 15-l: This list is acknowledged as a workable, feasible approach to quality patient care.

espousing safety), the EMS organization may be unable to defend that action if no written rules exist. Although safety is widely accepted as the number one priority for every EMS provider, without the administrative backup of written guidelines, actions taken in the interest of personal safety may be difficult to defend legally.

When writing, reviewing, revising and implementing policies and procedures in an organization, it helps to gather the appropriate people into a committee. Workers of the 1990s often expect to participate in creating the rules that affect them and are less tolerant of autocratic heavy-handedness than previous generations. When formulating safety rules and standards, employee input is important not only so that a comprehensive program will be created, but also so that the concept of safety as an organizational priority will be promoted among peer field providers.

Two areas of concern for the committee might be, for example, to establish a policy regarding who is to be in charge of a prehospital scene through use of the Incident Management System model (see Chapter 6). Guidelines can then be established regarding who is to assume the duties of the incident safety officer as well as when such a position needs to be filled at an incident. Because pro-active safety measures can seem so nebulous, the presence of a person titled "Safety Officer" will promote awareness of the priority of safety on significant calls; this will rapidly filter down to an awareness of this on every EMS call. The Safety Officer need not (usually cannot, in fact) always be the same individual; on normal, uncomplicated scenes, one of the crew of two or three can assume the role of safety officer, in addition to other tasks. Larger EMS services may have a designated safety officer who handles the administrative aspects of the safety program. Since one person cannot respond to all calls, other providers may be trained to serve as incident safety officers. *Every* EMS provider must be constantly reminded that the goals of on-scene safety fall to each of them continuously.

One volunteer EMS organization identifies the safety officer on a call-to-call basis, depending on who responds; that person is easily identified as the one holding the fire extinguisher. On situations of larger scale, one person should do only tasks related to the role of safety officer. This person must remain outside the stream of hands-on care in order to maintain a wide-angle view of the scene and the safety of those present. Once field personnel realize that a safety officer will be dependably present, they will know that safety matters at all times. (More information on the Safety Officer is presented in Chapter 16.)

Another way to impress upon field personnel that issues of safety are an organizational priority is to make safety information part of the study material for promotion.

One common trait among rescue personnel is to take risks on behalf of those they serve. Written policies and procedures will not replace this habit overnight. The manager must help generate a shift in the organizations environment. Personnel must understand clearly that being a professional should mean doing what is right for the patient **in the context of what is safe for the rescuer.** In an era where EMS personnel may be required, for example, to wait an hour for a properly equipped and trained hazardous materials team to respond when patients in the hot zone are clearly in need of help, written policies and procedures should help.

Suggestions to those in the process of clarifying the organizations commitment to safety must address the following general areas of concern:

• Approaches to roadway incidents, when and how to use reflective cones or flares, parking decisions, general driving safety and low force driving (see Chapter 2).

- Emergency vehicle design, which should include: passive restraint systems (airbags) in the cab; harness restraints or similar systems in the patient compartment; built-in compartments for maps, books, clipboards, etc.; and other safety design considerations (see Chapter 2).

- Appropriate use-and non-use-of lights and siren (see Chapters 2 and 4).

- Weight limits for safe lifting. These should include the weight of the patient plus the equipment being used, such as the stretcher, oxygen and EKG monitor. Provide alternative solutions, such as asking first responders to help or summoning a second EMS crew (see Chapter 3).

- Use of personal protective gear (see Chapter 5).

- Appropriate decision-making regarding leaving dangerous scenes either prior to contacting the sick or injured, or after initiating care. Also, which calls will require the presence of law enforcement prior to EMS entry (see Chapters 6, 7, and 8).

- Appropriate handling of the EMS-related aspects of hazardous materials incidents (see Chapter 9).

- Helicopter safety (see Chapter 10).

- General health and fitness programs and appropriate maternity and pregnancy policies (see Chapter 13).

- Required actions for minimizing the potential for contracting infectious disease (see Chapter 14).

This list is just a starting point. Each organization must consider its existing safety program and how best to bring the concepts involved to the proper level of priority.

Good Interagency Relations

Professionals from the other emergency agencies have tasks related to prehospital scenes. Sometimes these overlap or conflict with the EMS providers needs. For example, at traffic crashes, law enforcement personnel are often most interested in keeping traffic moving. Fire suppression personnel may be worried about the nearby overturned load of pesticides. Overall site safety of all emergency personnel must be universally acknowledged as the foremost priority. This is far easier to control and promote when interagency relationships are positive than when rivalry and turf battles interfere with everyone's mission to care for people in unfortunate circumstances.

Good interagency relations at EMS scenes should be available to crews at *every* scene, even those with a small number of personnel. Then, when such events such as mass casualty incidents needing mutual aid occur, scene safety and effectiveness are amplified by the cordial and familiar relationships that already exist. An atmosphere of cooperation will be easy to attain. The goal of good interagency relations is to foster a variety of quick, helpful backups-from police and hazardous materials team responses to fire suppression or first responder assistance.

It helps to have police, fire, rescue and EMS personnel train together so that the same terms mean the same thing to people from the different agencies (such as "staging area" and "triage"). Where relationships are good, maintaining them may be as easy as hosting an annual picnic or a similar event. If interagency relationships are strained, change will require extra time and effort initially-but it is an attainable goal. One service persisted for six years before reversing terrible relations with the first responders, so that now they work well together.

Improvement of interagency relationships is especially possible when the effort is initiated by top-level managers. Prehospital scenes are *not* the place to resolve or even discuss problems between organizations. Anecdotes abound of such incidents-many of which resulted in harm. In one case, two paramedics were hit by a motor vehicle at the scene of a roadway incident because a California Highway Patrol officer allegedly would not allow them to put out flares. They were both killed. In another case, which was more detrimental to the patient than the responders, a scuffle took place between firefighters and police officers over who was going to extricate a person from a crashed car. This type of conflict should never be an issue at the scene. Teamwork is the key to a safe, successful program.

Of the EMS team, one team of players who are not physically at the emergency scene, but who are central to the success of any mission are the professionals in the dispatch center. EMD-emergency medical dispatch-is a priority-setting, public-assisting concept which every manager interested in staying abreast of national standards for emergency care must investigate. Implementation of a proper EMD program has numerous safety advantages, which are outlined in Chapter 4. Proper implementation can mean significant cost savings. Nurturing this relationship (which is often an interagency situation) is also worthwhile.

Summary

Managers must implement measures to make the EMS workplace as safe as possible-on physical, emotional and mental terms. To fail to do so is to court lawsuits from all sides: from employees, patients, families, and the general public. Safety awareness demonstrates appropriate concern for the well-being of those working for the organization. The effect on morale and attrition-in themselves vigorous promoters of safety when the first is high and the second is low -cannot be disregarded. The way the manager implements safety as a top priority also affect whether a program is to succeed, both within and outside the organization.

References/Endnotes

1. Casey Terribilini and Kate Democoeur, "Pre-employment Back Injury Screening: The Employer's Crystal Ball,"JEMS, October 1991.
2. "Inside EMS: Buckle Up or Pay Up," JEMS, April 1985.
3. Charles Clark, "The Bottom Line," Azstar Safety Net (newsletter), October 1991.

CHAPTER 16

Building a Focus on Safety

Chapter Overview: In this chapter are found descriptions of the role and responsibilities of the Safety Officer. This is an increasingly common position-paid or volunteer-which can benefit the organization both in financial terms and in boosted employee morale. Many EMS organizations even empower a Safety Committee to oversee safety-related programs and tasks. Closely tied to a comprehensive safety program is the gathering of data and statistics to validate and measure the impact of various programs.

This chapter is designed to encourage high-level leaders in EMS organizations to build a focus on safety. With foresight, planning, funding and follow-through, safety in the workplace becomes less of a gamble and more of a reasonable expectation.

The way things are done at any particular organization is largely based on the unique culture it evolves over time. Although rules can be written and learned, and supervisors can try to enforce them, the unwritten rules in an organization often dictate which ones are followed most genuinely and consistently. (For example, for many years the culture within the fire service applauded "smoke eaters" rather than those who followed the rules and wore all the issued personal protective gear.) Knowing this, managers can implement programs for safety in ways that will use the existing corporate culture to guarantee success.

The organization has a responsibility to nurture an internal culture that supports an attitude of respect for the importance of safety. The organization has to support people who demonstrate adherence to the principles of safe practice of EMS. Both the written rules *and* the, corporate culture from which the unwritten rules are derived must reflect the importance of safety. Prehospital professionals need (and deserve) as stable an environment in which to conduct their work as possible.

In addition, the individual has a responsibility to the organization. People tend to view their EMS agencies as permanent entities. Yet businesses come and go. When a team of people truly cares about the success of its organization, it will understand the relevance of promoting safety as a primary mission, both for themselves and for the enduring success of the organization.

The Safety Officer

In the late 1980s, the fire service began to champion the position of Safety Officer. Many EMS organizations also recognize the wisdom of this idea. Just staying abreast of the federal and local mandates to promote worker safety in terms of hazardous materials and infection control is daunting; add the many other aspects of safety addressed in this manual, and the role can easily be a full-time job. Even though safety is everyone's responsibility, the Safety Officer holds a key to reducing injuries and death in the EMS workplace. Thousands-even hundreds of thousands-of dollars stand to be saved through the effective implementation of this position.

Models from the fire service for both basic safety standards and the role of Safety Officer can be found in the NFPA's documents *Standard for Fire Department Occupational Safety and Health Program (NFPA 1500) and Standard for Fire Department Safety Officer* (NFPA 1521-1992 edition). A person in this role is responsible for knowing:

- current local, state and federal laws pertaining to occupational safety and health that apply to EMS professionals.
- occupational hazards that exist-both actual and potential.
- factors related to physical and mental fitness and basic health,

such as basic knowledge of exercise physiology, good nutrition and stress management.

- how to properly and effectively manage a program designed for safety and health.

The Safety Officer has a job that requires appropriate experience with and knowledge of EMS operations. It demands a certain degree of literacy, ability to work with data and numbers for record-keeping purposes, skills in personnel management and interaction and a knack for interagency relations,

A Safety Officer can be part of any size or type of EMS organization. Small, volunteer and/or rural services may rotate the administrative Safety Officer position and designate the incident Safety Officer on a per-scene basis. Medium-sized organizations may delegate Safety Officer functions to the training officer, since there are many crossover tasks. A large organization can benefit tremendously by having a full-time Safety Officer.

An important element of the Safety Officer's role is to be able to report directly to whoever is the highest ranking leader of the organization. With so much responsibility for the well-being of personnel, the Safety Officer's ability to do the job must not be hampered by others along the chain of command; this could compromise the mission if health or safety issues arise in the areas supervised by resentful or suspicious intermediate managers.

Roles and tasks of the Safety Officer consist of both field operations and administrative functions. It is impossible for one person to be present at each EMS call even in the smallest organizations; therefore, one ongoing educational task is to permeate crews with a focus on safety. Safe practice must be continually sought by each member of the organization. Additionally, the Safety Officer can or should:

- respond-or send a trained designee-to major incidents or those which involve or may involve unusual safety hazards. At the scene, the Safety Officer *must not* engage in patient care or triage activities but instead should remain an observer. This person's job is to notice (and mitigate, where possible) hazardous or careless actions and to assess for changing circumstances (such as the start of a fuel leak or the onset of rush hour on a roadway) that could increase the existing hazard level. The Safety Officer should report to the Incident Commander upon,, arriving at the site. These two people should work together closely on behalf of both the field providers and the patients.

- assess personnel at EMS scenes-especially those of prolonged duration-for those needing on-site rehabilitation. Even temporary relief of duties allows for rest and rehydration (see Chapter 12). The Safety Officer can arrange for a resting site away from the action and perhaps can arrange for an EMS crew and the local Critical Incident Stress Management Team to staff it.

- attend post-incident critiques, adding positive and negative reviews of the safety angle where appropriate.

• identify and correct safety hazards in the workplace. This includes hazards in the organization's offices, training sites, stations, vehicle repair shops, supply rooms and so on. The Safety Officer has responsibility not only for field providers but for support staff and office workers as well.

• know the applicable laws, regulations and requirements as written by federal and state governmental bodies (such as OSHA) and national standards-making groups (such as ASTM and NFPA). Routinely assess departmental policies, procedures and operations for compliance. When compliance is not achieved, members of the organization need to be educated about the risks: there may be large fines or other penalties. Worse, someone could be seriously injured or killed.

• investigate all on-the-job injuries, illnesses and exposures (both hazardous materials and infectious diseases). Appropriately document the event and take (or recommend) appropriate follow-up action to minimize the chances that similar events could occur again.

• keep accurate personnel records. This includes a confidential file on each employee concerning occupational injuries or illnesses, as well as medical testing and treatment for them. These records are vital to support worker compensation claims. In addition, the Safety Officer should have records of routine health and medical examinations, documentation of physical fitness training requirements and achievement and exposures to infectious disease or hazardous materials (with follow-up actions).

• maintain records of inspections and service records on equipment and facilities used by the organization. This includes items such as seldom-used personal protective equipment and monitor-defibrillators. The Safety Officer should know what records are being kept by the vehicle maintenance department and whether those records are thorough and accurate.

• investigate emergency vehicle crashes and keep records of follow-up information. Substance-abuse testing may sometimes be needed. The Safety Officer must know the correct procedures to follow.

• keep records of situations that pose potential liability to the organization. These may include threats to sue, actual lawsuits and other service complaints.

• act as the safety liaison to other agencies. In addition, the Safety Officer should be the company's liaison with equipment manufacturers (including vehicles), standard-setting authorities (such as protocol committees) and other regulatory agencies (such as OSHA and in-state equivalents). By doing so, the Safety Officer can maintain a global view of the safety issues faced by the organization and make appropriate recommendations to the Safety Committee and the head of the organization.

• act as a two-way information resource. Listen to the suggestions and recommendations of all employees and introduce them to management. Relay new information and programs from managers or outside regulatory agencies to field personnel.

The Safety Officer must have appropriate functional authority. It does no good to create the position of Safety Officer without empowering this individual to make difficult, and sometimes unpopular, changes on behalf of safety, Without top-level administrative support (both verbal and financial), the task may sometimes be unreasonably troublesome or even impossible.

The Safety Committee

With so many responsibilities, the Safety Officer in some organizations will achieve faster and possibly better results with a properly formed and management-endorsed Safety Committee. In the spirit of collaborative or participative management, this group can assist the Safety Officer to identify and eliminate or minimize organizational safety issues. How an EMS organization does this may vary, with smaller ones delegating or sharing the various functions.

People to consider for such a committee include:
- the Safety Officer.
- the Medical Director.
- the Quality Improvement or Quality Assurance Coordinator.
- well-respected field personnel. (Without this group, the safety program faces an uphill battle in achieving credibility among the main group being protected.)
- a representative from the communications center.
- appropriate members of the management team.

The committee must be empowered to write and help implement safety programs, or else its existence may be viewed as being lip service to a topic of serious importance to the people it is designed to help protect.

The Safety Committee can be an active and busy group, and it must address various angles related to safety:
- laws, regulations and procedures.
- equipment.
- specific incidents (including crash-review board).
- recommendations for disciplinary action.
- specialized knowledge about infection control, hazardous materials and technical rescue operations.
- training and education.
- liaison activities with other agencies and organizations.
- data collection and information management.

Although the Safety Committee is an internal group intended to address safety and health issues, it may elect to generate positive public relations (to the staff or even to the general community) by working with those responsible for public education and information. For internal public relations (PR), the message is that a safe work place helps personnel be better prepared to attend to their patients. For external PR, such efforts can bolster that precious and intangible commodity of public trust. This can help foster cooperation by members of the public with field providers, which is a safer environment for them. Shattered public trust might happen, for example, through the misdeeds of even just one team of overstressed EMS personnel; high stress levels can be readily related to unsafe actions (see Chapters 4 and 12).

The work of an effective safety committee can help reduce costs to the organization. Cost savings can be realized regarding insurance premiums, fewer crashes, fewer collision-related lawsuits, reduced wear and tear on equipment, diminished attrition related to injury (or fear of it), costs of covering injured personnel (plus the expenses related to those injuries, such as paying the people who are working overtime) and so on. The direct benefit to everyone in the organization is a healthier bottom line for the organization-some of which might be applied toward bonuses or raises in pay.

Safety-Related Data and Statistics

Data collection about safety issues is a pressing need in EMS. Reasons for collecting data are to focus on trends, to design relevant training programs, to evaluate equipment and to establish policies to reduce injuries and illnesses. But adequate data collecting is almost non-existent. Where it exists, it is not standardized from region to region. Data collection could go far to help cap avoidable expenditures in EMS.

No one really knows how many EMTs and paramedics are injured in a given year, or how those injuries occurred. No one can say authoritatively how many people died, or how many days of work were lost for a back problem, or how many had pain in a knee or shoulder. Cost estimates are in the millions-but no one really knows for sure.

Data collection can help harried EMS managers see the result of a proper safety program. Good data can earn government dollars, lower insurance costs and produce other significant cost savings. It can help the development of suitable loss control and risk management programs for the EMS industry. Collecting generalized data on injuries and work-related illnesses is also an OSHA requirement; unfortunately, the amount of required data alone falls short of being able to identify useful trends.

Other Related Programs

An organization building a focus on safety may implement various other programs, possibly including:
- periodic health assessment (pre-employment and also periodically during employment)
- physical fitness programs
- stress management programs
- member assistance programs
- comprehensive infection control programs
- hearing conservation programs
- driver safety and training programs

- training and continuing education programs that stress safety
- incident management system that include emergency incident rehabilitation procedures.

None of these programs needs to feel like a burden to an organization committed to safety; many are likely to already be in place or relatively easy and inexpensive to implement. Each is addressed in NFPA 1500, which can be used as a guideline for setting up an EMS Occupational Safety and Health Program.[1]

In order to begin implementation of various programs, one must understand how each part of the puzzle contributes to the whole picture of EMS safety.

Summary

Building a focus on safety won't happen overnight, even in the best EMS organizations. To make the program work (and to realize the resultant savings), EMS managers must be committed to the concept. Without high level support, efforts by lower-ranking members of an organization can often become meaningless. Safety-related programs also depend on proper funding and staffing. The Safety Officer must be empowered to solve problems efficiently.

The long-overdue era of a safer work setting for EMS personnel is finally asserting itself. Anything the manager or owner can do to help implement this new standard of care must be initiated now.

References/Endnotes

1. Gordon Sachs, presented in an Emergency Education Network Videoconference, "Current Trends in EMS Management," Emmitsburg, MD: November 18, 1993.

* U.S. GOVEANMENT PRINTING OFFICE: 1994- 5 1 9 - 5 9 0 / 8 1 0 3 3

  Federal Emergency Management Agency

United States Fire Administration

FA-144/April 1994